Lecture Notes in Mechanical Engineering

Series Editors

Fakher Chaari, National School of Engineers, University of Sfax, Sfax, Tunisia

Francesco Gherardini, Dipartimento di Ingegneria "Enzo Ferrari", Università di Modena e Reggio Emilia, Modena, Italy

Vitalii Ivanov, Department of Manufacturing Engineering, Machines and Tools, Sumy State University, Sumy, Ukraine

Editorial Board

Francisco Cavas-Martínez, Departamento de Estructuras, Construcción y Expresión Gráfica Universidad Politécnica de Cartagena, Cartagena, Murcia, Spain

Francesca di Mare, Institute of Energy Technology, Ruhr-Universität Bochum, Bochum, Nordrhein-Westfalen, Germany

Mohamed Haddar, National School of Engineers of Sfax (ENIS), Sfax, Tunisia

Young W. Kwon, Department of Manufacturing Engineering and Aerospace Engineering, Graduate School of Engineering and Applied Science, Monterey, CA, USA

Justyna Trojanowska, Poznan University of Technology, Poznan, Poland

AF817361

Lecture Notes in Mechanical Engineering (LNME) publishes the latest developments in Mechanical Engineering—quickly, informally and with high quality. Original research reported in proceedings and post-proceedings represents the core of LNME. Volumes published in LNME embrace all aspects, subfields and new challenges of mechanical engineering.

To submit a proposal or request further information, please contact the Springer Editor of your location:

Europe, USA, Africa: Leontina Di Cecco at Leontina.dicecco@springer.com
China: Ella Zhang at ella.zhang@springer.com
India: Priya Vyas at priya.vyas@springer.com
Rest of Asia, Australia, New Zealand: Swati Meherishi at swati.meherishi@springer.com
Topics in the series include:

- Engineering Design
- Machinery and Machine Elements
- Automotive Engineering
- Engine Technology
- Aerospace Technology and Astronautics
- Nanotechnology and Microengineering
- MEMS
- Theoretical and Applied Mechanics
- Dynamical Systems, Control
- Fluid Mechanics
- Engineering Thermodynamics, Heat and Mass Transfer
- Manufacturing
- Precision Engineering, Instrumentation, Measurement
- Materials Engineering
- Tribology and Surface Technology

Indexed by SCOPUS and EI Compendex. All books published in the series are submitted for consideration in Web of Science.

To submit a proposal for a monograph, please check our Springer Tracts in Mechanical Engineering at https://link.springer.com/bookseries/11693.

R. S. Priyadarsini · T. Sundararajan
Editors

Advances in Small Satellite Technologies

Proceedings of National Conference on Small Satellite Technology and Applications, NCSSTA 2020

Springer

Editors
R. S. Priyadarsini
Department of Civil Engineering
College of Engineering Trivandrum
Thiruvananthapuram, Kerala, India

T. Sundararajan
Space Transportation Systems
Vikram Sarabhai Space Centre
ISRO
Thiruvananthapuram, Kerala, India

ISSN 2195-4356 ISSN 2195-4364 (electronic)
Lecture Notes in Mechanical Engineering
ISBN 978-981-19-7473-1 ISBN 978-981-19-7474-8 (eBook)
https://doi.org/10.1007/978-981-19-7474-8

© The Editor(s) (if applicable) and The Author(s), under exclusive license to Springer Nature Singapore Pte Ltd. 2023
This work is subject to copyright. All rights are solely and exclusively licensed by the Publisher, whether the whole or part of the material is concerned, specifically the rights of translation, reprinting, reuse of illustrations, recitation, broadcasting, reproduction on microfilms or in any other physical way, and transmission or information storage and retrieval, electronic adaptation, computer software, or by similar or dissimilar methodology now known or hereafter developed.
The use of general descriptive names, registered names, trademarks, service marks, etc. in this publication does not imply, even in the absence of a specific statement, that such names are exempt from the relevant protective laws and regulations and therefore free for general use.
The publisher, the authors, and the editors are safe to assume that the advice and information in this book are believed to be true and accurate at the date of publication. Neither the publisher nor the authors or the editors give a warranty, expressed or implied, with respect to the material contained herein or for any errors or omissions that may have been made. The publisher remains neutral with regard to jurisdictional claims in published maps and institutional affiliations.

This Springer imprint is published by the registered company Springer Nature Singapore Pte Ltd.
The registered company address is: 152 Beach Road, #21-01/04 Gateway East, Singapore 189721, Singapore

Conference Committee

Chief Patron

Prof. (Dr.) Byju Bai T. P., Director of Technical Education, Government of Kerala

National Advisory Committee

Shri. Somanath S., Director, VSSC, Indian Space Research Organisation, Trivandrum, India
Shri. P. Kunhikrishnan, Director, U. R. Rao Satellite Centre, Bangalore, India
Shri. D. Sam Dayala Dev, Director, IISU, Indian Space Research Organisation, Trivandrum, India
Dr. V. Narayanan, Director, LPSC, Indian Space Research Organisation, Trivandrum, India
Shri. A. Rajarajan, Director, SHAR, Indian Space Research Organisation, Trivandrum, India
Shri. Rakesh S., Chairman, Antrix Corporation Ltd., India
Shri. Vinay Kumar Dadhwal, Director, Indian Institute of Space Science and Technology, Trivandrum, India
Dr. A. Ajayaghosh, Director, NIIST, Trivandrum, India
Dr. V. Koteswara Rao, President, Society for Small Satellite Systems, India
Dr. Anil Kumar, Director, DSSAM, Indian Space Research Organisation, India
Shri. Kris Nair, CEO, KAWA Space, India

General Chairs

Prof. (Dr.) Jiji C. V., Principal, College of Engineering Trivandrum, India
Shri. D. V. A. Raghava Murthy, Secretary, Society for Small Satellite Systems, India
Shri. P. S. R. S. Sastry, Director, DSP, DRDO, India

Conference Chairs

Prof. (Dr.) Sindhu G., Dean (R), College of Engineering Trivandrum, India
Prof. (Dr.) Jayakumar S., Dean (PG), College of Engineering Trivandrum, India
Prof. (Dr.) Abdul Nizar, Dean (UG), College of Engineering Trivandrum, India

Organizing Secretary

Dr. R. S. Priyadarsini, Professor in CE, College of Engineering Trivandrum, India

Joint Secretary

Dr. Ranjith S. Kumar, Associate Professor in ME, College of Engineering Trivandrum, India

Treasurer

Dr. Mabel Ebenezer, Associate Professor in EE, College of Engineering Trivandrum, India

Finance and Accounts Committee

Dr. Salini S., Associate Professor in CE, College of Engineering Trivandrum, India
Dr. Sreelatha G., Associate Professor in EC, College of Engineering Trivandrum, India
Dr. Mini R. S., Associate Professor in ME, College of Engineering Trivandrum, India

Peer-Review and Publication Committee

Dr. Abhilash Suryan, Associate Professor in ME, College of Engineering Trivandrum, India
Dr. T. Rajeev, Professor, Associate Professor in EE, College of Engineering Trivandrum, India
Dr. Rajeev Rajan, Associate Professor in EC, College of Engineering Trivandrum, India

Technical Programme Committee

Dr. Dinesh Gopinath, Associate Professor in EE, College of Engineering Trivandrum, India
Shri. Ajith R. R., Associate Professor in EE, College of Engineering Trivandrum, India
Dr. Sreeni K. G., Associate Professor in EC, College of Engineering Trivandrum, India
Shri. Sharafudeen, Computer Programmer in the Department of CS, College of Engineering Trivandrum, India

Registration and Correspondence Committee

Dr. Sreeja S., APEE, Associate Professor in EE, College of Engineering Trivandrum, India
Smt. Gigy P. G., Associate Professor in EE, College of Engineering Trivandrum, India

Publicity Committee

Shri. Sajeev Mohan (Coordinator), Associate Professor in Physics, College of Engineering Trivandrum, India
Shri. Praveen A., Associate Professor in ME, College of Engineering Trivandrum, India
Dr. Sini V. Pillai, Associate Professor in MBA, College of Engineering Trivandrum, India

Website Designing

Shri. Vipin Vasu, Associate Professor in CS, College of Engineering Trivandrum, India

Preface

The 2nd *National Conference on Small Satellite Technology and Applications* (NCSSTA 2020) was organized and hosted by College of Engineering Trivandrum (CET) in collaboration with Society for Small Satellite Systems (SSSS) during December 11–12, 2020. This two-day event sponsored by All India Council for Technical Education (AICTE), Government of India, was conducted as a virtual conference as a part of the 80th year celebrations of College of Engineering Trivandrum.

Small satellites are becoming increasingly popular in numerous applications that were accomplished earlier with medium or large satellites. Applications such as high bandwidth internet, high-resolution images, asset tracking, disaster management, agriculture, water resources, geology and urban planning now employ the services of small satellites. The rapid rise in demand has called for better technology in miniaturization, design, production, data acquisition and storage. In India, there is considerable demand for satellite-based services for communication purposes seeking an increase with implementation of initiatives like 5G and Internet of Things. Many companies across the world are also coming forward to make use of this huge market by developing small satellite launch vehicles. This conference has provided a platform for the technologists and students working in these domains to showcase their research outcomes and interact with the pioneers and eminent personalities in this field and get wide exposure.

The catalytic role provided by the conference resulted in rich, purposeful and engaging deliberations among students, technologists and scientists. During the two-day program, 4 plenary sessions were held, along with 3 keynote addresses by eminent personalities. The success of the event was due to the sincere patronage and participation by the delegates throughout the event, who numbered about 90. Technical sessions focused on payload design, remote sensing, orbital platform for onboard experiments, systems, mechanical design and navigation.

Proceedings of NCSSTA-20 are a compendium of technical papers submitted by various research groups of our country. A total of 48 papers were selected and subjected to two-tire review processes by experts from various industries and research institutions. Out of these, 36 contributory papers from various government and private organizations like VSSC, IISU, LPSC, LEOS and UR Rao Satellite Centre of ISRO, RIC IIT Madras research park, Student satellite team of IIT Bombay, Centum electronics, Somaiya Institute of Engineering and Information Mumbai, TSC Technologies, Hyspace Technologies, College of Engineering Trivandrum, Indian Institute of Space Science and Technology and TKM College of Engineering were selected for presentation in the conference. A chosen set of 26 from the 36 presented papers after a series of tough review exercises was further recommended for publication in the Springer proceedings. The editors gratefully acknowledge the patience shown by the reviewers for giving their valuable time for critical reviews that positively impacted the quality of the compendium. Care was taken to check plagiarism by adopting standard procedures and software tools. The editors are confident that the high-quality contents in the proceedings will benefit academic, research and industrial subscribers.

The editors place on records the contributions made by authors and co-authors for their contributions in the small satellite technology domain in general and for creating quality research papers for NCSSTA-20 in particular. The submissions were considered after disseminating the information regarding the Springer policies, and the editors assume that the authors are responsible researchers, and the contents reported are original and are reporting for the first time. However, the editors are not responsible for any scientific observations made by the authors, and any query regarding the technical correctness may be addressed directly to the authors. Owing to the tight time schedules followed for the conference and proceedings preparation, any unintended errors may kindly be absolved.

Thiruvananthapuram, India

Dr. R. S. Priyadarsini
Dr. T. Sundararajan

Contents

Orbital Platform and Payload Experiments

Systems for Small Satellites

Mechanical System Design for Small Satellites

About the Editors

Dr. R. S. Priyadarsini had her Engineering Education at the University of Kerala and at IIT Madras. Soon after PG studies she joined at IIT Madras in 1999 as Senior Project Officer and worked in the development of IS 800:2007 and also in developing the teaching and learning materials for Steel-Concrete Composite Structures. Later in 2000 joined as Assistant Professor at the University of Kerala. She taught at College of Engineering Trivandrum and Rajiv Gandhi Institute of Technology, Kottayam. She is currently working as Professor in the Department of Civil Engineering at the College of Engineering Trivandrum, Kerala. She is the recipient of Shree Gaythree Devi Award with Gold Medal from IIT Madras in the year 2012. Her major research interests include Advanced Composite Structures, Steel—Concrete Composite structures, Steel and RC Structures, Structural Stability, Smart Structures and Numerical modeling. She has more than 30 publications in her credit including National and International Journals and conferences. She is a member in many Professional bodies such as Institution of Engineers India, Aeronautical Society of India, ISTE, ISAMPE and Society for Small Satellite Systems.

Dr. T. Sundararajan is an expert aerospace structural dynamics professional, having three decades of experience at Vikram Sarabhai Space Centre (VSSC), Indian Space Research Organisation. As Associate Programme Director of Space Transportation Systems, he leads a resourceful team of young and experienced engineers to deliver domain specific solutions for various needs of India's launch vehicles. He earned his Ph.D. from Faculty of Engineering and Technology, University of Kerala. Earlier in his career, he groomed an ace team of engineers for dynamic characterization of ongoing and future aerospace systems/ launch vehicles. His penchant to adopt state-of-the-art developments in science/technology and update professional knowledge, made him foray into the realm of AI and ML for structural dynamic characterization/ model updating and Operation Modal Analysis. He has immensely contributed in enhancing indigenous finite element method based software by way of pan national collaboration with leading national research and academic institutions. Despite the demanding professional schedule requirements he carves-out time to engage in academic guiding and mentoring of Master's and Doctoral degree scholars. He has more than 40 peer reviewed journal and conference papers to his credit.

Payload Design for Small Satellites

Designing BeliefSat-1: An Open-Source Technology Demonstrator PocketQube

Rohit Bokade, Umesh Shinde, Supriya Bhide, Nikhil Shinde, Jyoti Gadgil, Manas Sable, and Rohan Gamare

Abstract BeliefSat-1 is the first student satellite of K. J. Somaiya Institute of Engineering and Information Technology, Sion, Mumbai being developed as a sub-part of the institute's proposed payload under ISRO's PS4OP programme. It is being developed as per PocketQube standard with 2P ($50 \times 50 \times 114$ mm^3, ≤ 500 g) unit specifications for satisfying mechanical, power and tracking requirements. The major goals of the satellite include the demonstration of innovative fabrication approach based on interlocked PCBs serving as a structural material, providing digipeater service to amateur radio community, flight proving communication and power system based on COTS electronic for usage in future missions. The design has been evolved around the Arduino ecosystem, to make it easily adaptable by future missions. The paper first highlights certain calculations for getting a range of operational parameters and then describes different subsystems implemented within the satellite design. The mission assumes 550 km sun-synchronous orbit. Overall, it uses Atmega1284p as on-board computer, IXYS monocrystalline solar cells for power generation, SPV1040 for maximum power point tracking, LORA1268F30 as transceiver, Arducam 2MP camera as imaging payload, Samsung 18,650 batteries

R. Bokade (✉) · U. Shinde · S. Bhide · N. Shinde · J. Gadgil · M. Sable · R. Gamare
K. J. Somaiya Institute of Engineering and Information Technology, University of Mumbai, Sion, Mumbai, India
e-mail: rohit.bokade@somaiya.edu

U. Shinde
e-mail: ushinde@somaiya.edu

S. Bhide
e-mail: supriya.bhide@somaiya.edu

N. Shinde
e-mail: shinde.nv@somaiya.edu

J. Gadgil
e-mail: jyoti.gadgil@somaiya.edu

M. Sable
e-mail: manas.sable@somaiya.edu

R. Gamare
e-mail: rohan.gamare@somaiya.edu

© The Author(s), under exclusive license to Springer Nature Singapore Pte Ltd. 2023
R.S. Priyadarshini and T. Sundararajan (eds.), *Advances in Small Satellite Technologies*, Lecture Notes in Mechanical Engineering, https://doi.org/10.1007/978-981-19-7474-8_1

and COTS sensors (HMC5883L magnetometer, MPU6050 gyro-sensor). The paper also describes minimal setups for enabling reception by students all over India.

Keywords PocketQubes · Arduino based pico-satellite · COTS for nano-satellite · FR4 structure for satellite · PS4-orbital platform · Open-source satellite · Amateur radio digipeater · SSDV image transmission

1 Introduction

ISRO's PS4-Orbital Platform [1] is an opportunity for academic institutions and research laboratories to carry out in-orbit experiments on the fourth stage of PSLV after deployment of the main payload. Student satellite team of K.J.S.I.E.I.T. has proposed an experiment for developing and demonstrating indigenous capabilities for enabling the adoption of [2]. PocketQube standard pico-satellites. The experiment involves an open-source deployer design from which a PocketQube standard satellite will be deployed to demonstrate deployment capabilities. BeliefSat-1 is the satellite that will be deployed out. As it is the first mission, the development approach is to use it to flight-prove the designs of basic subsystems, which can pave way for future payload intensive missions. If the proposed design succeeds, future PocketQube missions will be able to integrate the same COTS component-based power management, communication, structure and on-board computer subsystem designs with their payloads with required modifications, thus reducing the time and efforts required. PocketQube standard is an emerging standard especially preferable for student and technology demonstration missions requiring shorter development lifecycle and launch budget constraints. With the rise of attraction towards the space sector, a lot of student satellite programmes in India are also rising. BeliefSat-1 design is intended to be a good starting point. Along with the technology demonstration, the satellite will also need to have a secondary amateur radio payload as it will be using amateur band for the operation. The design will be open sourced on the project's [3] GitHub repo under CERN-OHL V1.2 for reference of future student satellite developers.

2 Experimental Procedure/Methodology/System Description

2.1 Constraint Analysis

Mechanical constraints

- The design should follow dimension and mass constraints specified by the PocketQube standards.

- Antennas should be in a folded state when inside the deployer and there should be a mechanism to deploy them only after 30 min from separation from the launch platform.
- The materials used in fabrication of the structure should be easily machinable in college premises or one should be able to order it machined in custom shape and sizes inexpensively.
- Material should have outgassing characteristics within limits specified in PQ standards.
- Components and parts should be COTS and having a flight history will be preferred.
- Components should preferably have community maintained open-source firmware libraries.

Orbit constraints

- The PS4OP programme is supposed to operate in the 500–700 km range of orbit height, so the system design should be designed based on orbit height in that range.
- Assuming 550 km sun-synchronous orbit, the free space path loss for an RF signal in UHF ham band is around 149 dBm when at 15° elevation (generally satellites are visible when above this elevation with respect to observer location in urban areas due to surrounding high rise buildings) with respect to the observer's horizon. Thus, transmitter power should satisfy this link budget.
- Orbital time period of this orbit is around 96 min, thus the power budget of the system will need to be designed to sustain the eclipsed part of the orbit.

2.2 *Payload Selection*

1. Amateur radio payload: As per the core goal of the mission of promoting amateur radio activities in India and driving student interest towards it, an amateur radio "store and forward digipeater" which should be specifically designed such that it could be triggered by inexpensive off-the-shelf radio modules, was selected as payload. This payload was also essential to fulfil the mandatory requirement for [4] IARU frequency coordination that any satellite using amateur radio bands for communication should have at least one amateur payload as primary or secondary functionality.
2. Technology demonstration of the novel structure fabrication approach, communication, power and on-board computer subsystem made from COTS electronics. This will also include further performance analysis of on-board general purpose COTS magnetometer and gyro-sensor so that their usefulness in future missions can be determined.
3. Imagery payload and extended mission: To demonstrate on-board computer's ability to operate uplink activated science/tech payloads for defined amount of time and at defined operation frequency, a [5] COTS RGB camera was selected

to be put on-board which can capture and send image in SSDV format when commanded or may operate in mode for automatic image capture and downlink after defined time interval when commanded to do so. As no special attitude determination and control system have been planned for the mission, these images will be taken in random orientations. Although these images will not make any special usage due to the randomness of orientation, they may serve as training data set for machine learning algorithms for future missions which may correlate ambient light sensor, solar panel power, gyro-sensor and magnetometer readings with the manually determined orientations of these images to help develop camera-based or COTS sensor-based attitude determination system in future missions.

2.3 Operating Modes of the Satellite

- Deployment mode: After separation from the deployer, the system will turn on but will remain idle for 15 min or more if specified by the launch provider, so as to avoid the interference with the launch vehicle and to avoid mechanical damage caused by forceful deployment of the satellite antenna. After this idle time, cord burning mechanism will start and burn the cord holding antennas in folded state, deploying the antenna.
- Safe mode: Satellite enters this mode, when the battery levels are less than 50%. In this mode, the frequency of telemetry transmission will be reduced from once every 30 s to once every minute, to save power. Also, the output power of the transmitter will be reduced to < 0.5 W to save the power. When in this mode, even if it receives the command for image capture and transmission, it will not respond to it as the image capture process has heavy power requirements.
- Normal mode: Satellite automatically enters this mode once the battery level crosses 70% and will stay in this mode until commanded to automatic imagery mode or till the battery levels fall below 50%.

3 Analysis

3.1 Power Subsystem

- IXYS SM141K06L [6] satisfied the requirement of being available off-the-shelf, while also being relatively inexpensive and generating a fair amount of power (184 mW/cell). Their dimensions allow 4 of them to be put on the longer side and 2 on the shorter side, while leaving space for putting on the sensors. They have a flight history aboard WREN which was one of the initial PocketQubes and is being used by other recent missions like [7] FossaSat-2.
- SPV1040 [8] has been used as MPPT IC in [7] FossaSat-1 and satisfies the requirements of MPPT on selected solar cells too.

18,650 batteries have a good flight history aboard CubeSat missions and thus are selected along with the metallic battery holders for the mission. Each solar panel has solar cells, ambient light sensor, current sensor, MPPT IC while also serving as a structural side element. This makes the design of panels modular and easy to adapt to in future missions. Interface is a 5 pin interface with power, ground, i2c lines and an enable pin.

3.2 On-Board Computer and Sensors

- Arduino has been chosen as the target platform, as a range of different microcontrollers, catering different requirements of different use cases are compatible with it. Numerous community driven and well tested libraries for range of sensors and applications along with a large number of tutorials, community support forums, easily and cheaply available hardware make it a good platform with gentle learning and adaptation curve.
- For BeliefSat-1 [9], Atmega1284p has been chosen with 8 MHz clock and 3.3 V io voltage implementation as it provided enough FLASH and SRAM especially for the large buffers used in SSDV packet framing [10] and Reed-Solomon forward error correction while having reasonably low power consumption of 4 mA (essential feature as MCU is the part that remains active all the time) in this configuration. Built-in watchdog useful for resetting the MCU if it gets stuck in a loop by radiation induced single event upset or single event latch-up and brown-out detection to stop the calculations when supply voltage fall below defined voltage so as to avoid faults in operation and damage to peripherals make it suitable for the application in hostile remote environment of space.
- Two [11] LM75 temperature sensors are implemented on the main board to monitor the temperature of the batteries which are held on both sides of PCB. [12] HMC5833L COTS magnetometer and [13] MPU6050 COTS gyro-sensor are also placed on the main board. Resistor divider circuit with two 10 k resistors, tied to analogue pins which are basically inputs to internal ADC of MCU are used to monitor the battery voltage. VEML7700 [14] ambient light sensor and INA219 current sensor are placed on the solar panels (Table 1).

3.3 Communication Subsystem Design

The [17] LORA1268F30 module by NicreRF based on Semetech's SX1268 RFIC has been selected as on-board transceiver owing to its support for wide range of bandwidth, data-rate and choice between both spread-spectrum-based Lora modulation and widely used (G) FSK modulation. A large 255 byte on-chip FIFO allows

Table 1 Rationale for selection of COTS sensors

COTS sensor	Rationale for selection
LM75	Has a measurement range of – 55 °C to + 125 °C with accuracy of ± 3 °C good enough for coarse measurements of interior temperature of the satellite. The range is enough as per observations in analysis of flight data [15] of Fossasat-1
HMC5833L	The sensor has a resolution of 5 milli-gauss, i.e. 0.005 gauss, which is enough to measure the variations in field strength of Earth's magnetic field, which is 2.0E−5T, i.e. 0.2 gauss at 600 km as per [16]
MPU6050	This is used to take gyro-sensor readings in the body centred frame, thus will be responsible for measuring the angular velocity over 3 axis. The data will be useful to evaluate the performance of detumbling arrangements over the time. The sensor is able to measure with high accuracy of 131 LSB/(°/s) at full scale range of ± 250°/s which is high enough to measure the tumble rates
VEML7700	The sensor has a range of 0 to 120,000 lx with resolution of 0.0036 lx. Although the author has not found any flight history for it, the sensor can be a good option as a coarse sun sensor in future flights once proven upon this
INA219	Has required sensing range and flight history aboard NTNUs CubeSat projects NUTS-1 [17]

easier implementation of large packets, e.g. in the case of SSDV packets. Ability to output up to 32.5 dBm of power makes it suitable for the required RF link budget. Built-in packet engine helps reduce compute requirements of MCU by automatically detecting and separating the preamble, sync-word and other parts of the packet.

3.4 Structural Design

FR4, the material used for printed circuit boards, has required strength, flight history aboard many space missions, is cheaply and widely available and can be cut into different sizes and shapes as per the requirement using minimal tools like table top CNC or PCB prototyping machines, likes of which are easily available in most colleges and hackerspaces. Apart from being good structural material, it can also be used to mount solar cells and sensors. Thus, an innovative design with interlocked PCBs held together by PCB stand-off screws and reinforced by orthogonal soldering pads was used. The structure is made within the PocketQube standards and implements 2 end-stop switches on bottom plate which remain closed when the satellite is inside the deployer, cutting the power to the system until it is deployed, so as to ensure there is no premature deployment of antennas which are made out of pieces of measuring tape, held in folded position by a nylon cord that is cut by heated resistor only when the system is turned on. The design of these interlocking PCBs was made by first developing a general prototype PCBs with slots and protrusion, which were then checked for fitting, flimsiness and ability to withstand large forces, while still

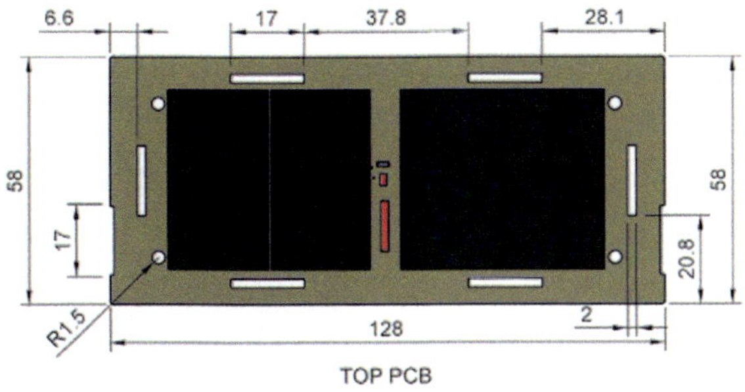

Fig. 1 Top PCB

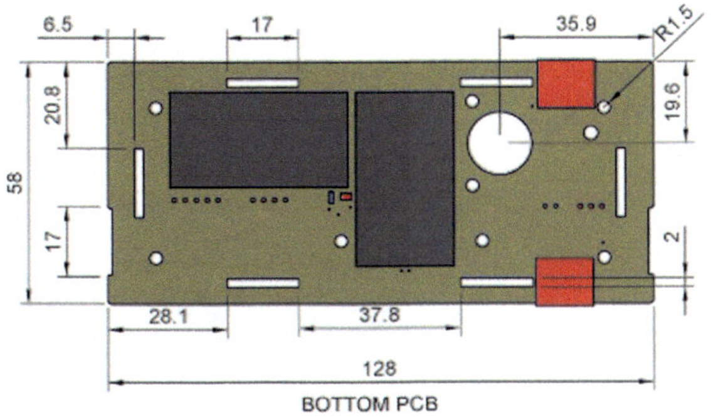

Fig. 2 Bottom PCB

maintaining the form factor. This was tested by a coarse method of dropping the assembled prototype from a height of 20 m on a concrete floor. After the drop, the displaced parts and overall form were analysed; it was observed that, though some parts of structure displaced from their position by angle of around 0.5–1°, the overall structure survived the drop. This analysis was further taken into account while developing the finalized CAD design where instead of single slot and protrusion, a pair of them were implemented on longer sides, which suffered minor deflection under drop. Later Fusion 360 design software's loading analysis and frequency modal analysis were used to simulate the load and vibrations on the proposed final structural design. Following are the dimensions finalized and simulation results are discussed in the results section (Figs. 1, 2, 3, 4, 5 and 6).

3.5 Antenna Orientation

As mentioned before (Sect. 2) one of the goals during the extended mission of the proposed satellite is to take images at random orientations while recording readings

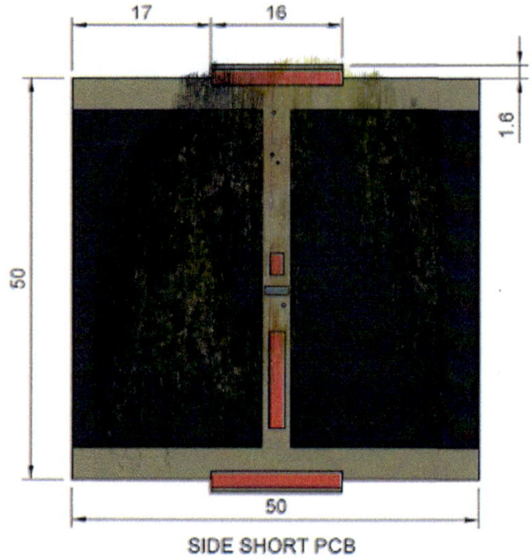

Fig. 3 Short side PCB

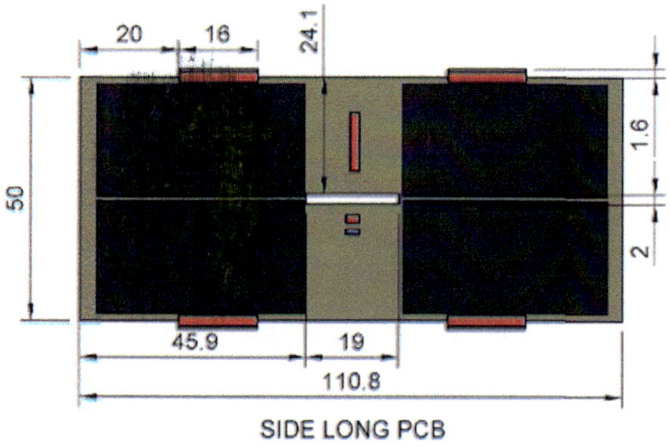

Fig. 4 Long side PCB

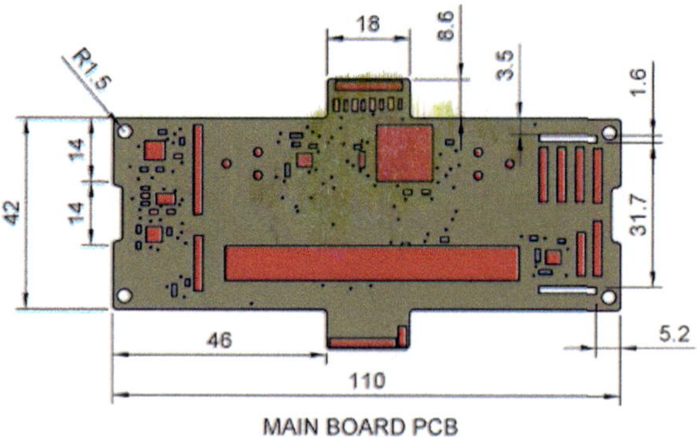

Fig. 5 Main board PCB

Fig. 6 Structural drop test prototype

of COTS sensors, thus there is no special requirement of pointing for imagery as the images are to be taken at random orientations. However, the communication link might suffer pointing losses or frequent signal fadings if the angular rate is not damped and antennas are not oriented within the directivity range. Thus, 2 slots in the main PCB were made where up to 4 neodymium N42 magnets (1/16″dia. × 1/2″thick with 1.32 T strength) can be fitted. These magnets were selected due to their small dimensions and mass, along with good field strength. Currently, the team is working on the amount of Mu-Metal sheet required for dampening the oscillations based on past designs used in Fossasat-1 and Colorado Student Space Weather Experiment. The goal of PMAC design currently being worked on is to keep the antennas within ± 25° of the local magnetic field within 14 days so as to keep the pointing losses of the antenna minimal. MU-metal sheet will be wrapped and held through the already present slots within the structure, and thus, no additional modification to structure is required after finalization. The antenna used is a tape measure dipole, which when simulated gives a radiation pattern as follows. The antenna has a simulated gain of 2.4 dB and half power beamwidth is around 78.8°. Thus assuming orbit height to be 550 km, the ground track would be 2 * 550 * tan((78.8/2)) = 903.55 km. The detailed design and simulation of the PMAC proposed is out of the scope of this paper (Figs. 7 and 8).

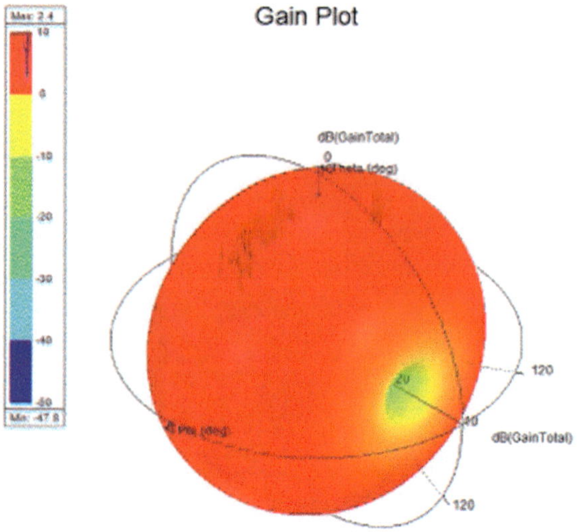

Fig. 7 Gain plot

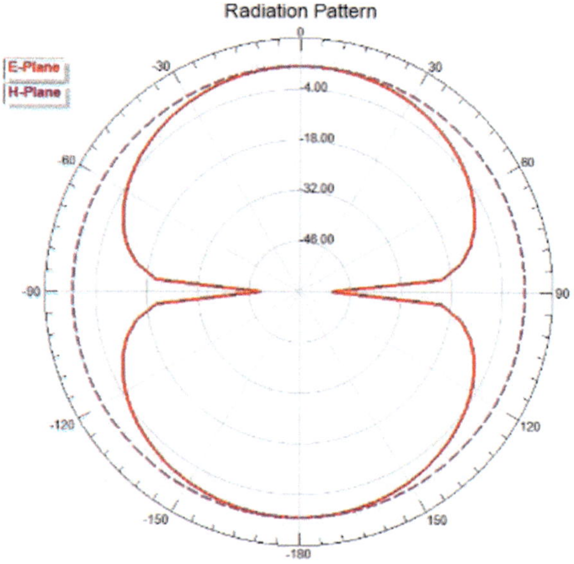

Fig. 8 Radiation pattern

3.6 Link Budget

See Table 2.

Table 2 Link budget

Parameter	Value	Reason/Remark
Assumed orbit height	550 km	PS4OP range 500–700
Minimum elevation above horizon for tracking	15°	High rise buildings in urban areas near home ground station cause obstruction at lower angles
Slant range at minimum elevation	1518.4 km	Used AMSAT link budget calculator [16] to calculate this
Frequency assumed	436.5 MHz	Midpoint of UHF ham band
Free space path loss at this frequency at the distance of slant range	148.9 dB	Used AMSAT link budget calculator to calculate this
Transmitter output power	29 dBm	The output power is < 30 dBm when voltage is lower than 4.5 V. This value is as per the datasheet of the module
Connector cable and mismatch losses	5 dB	Estimated
Satellite antenna gain	2.4	Simulated
Atmospheric losses	1.1	Used AMSAT link budget calculator to calculate this
Ionospheric losses	0.8	Used AMSAT link budget calculator for reference
Received power on Earth	− 125.4 dBm	Not including pointing losses and polarization losses
Gain of ground station antenna	11.8 dB	Theoretically calculated for 10 element Yagi-UDA antennas using DL6WU online calculator
LNA gain	22 dB	Using LNA4ALL COTS LNA in 70 cm
Losses due to connectors and cables at ground station	5 dB	Estimated
Filter insertion losses	5 dB	Chebyshev bandpass filter for 400–500 MHz
Gain of second stage LNA	22 dB	Using LNA4ALL COTS LNA in 70 cm
Received power at receiver	− 79.6 dBm	This excludes pointing losses and polarization losses
Receiver sensitivity at	− 115 dBm	Derived from sensitivities for other data rates given in the datasheet of transceiver for 9.6 Kbps FSK with 2.4 kHz deviation
Link margin	35.4 dBm	Enough to account for polarization and pointing losses

4 Results and Discussions

See Figs. 9, 10, 11, 12, 13, 14 and Tables 3, 4.

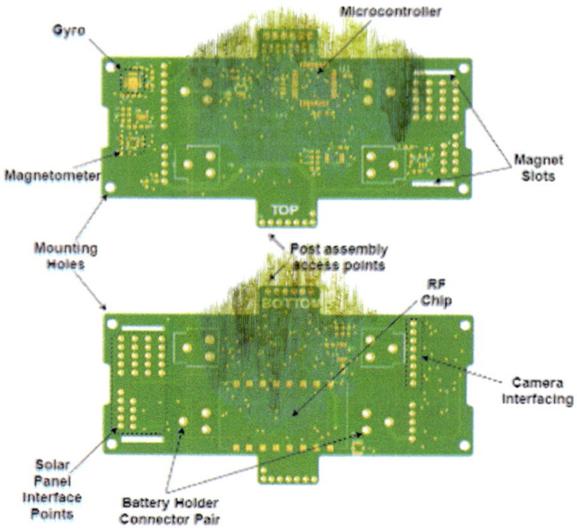

Fig. 9 Main PCB top and bottom side

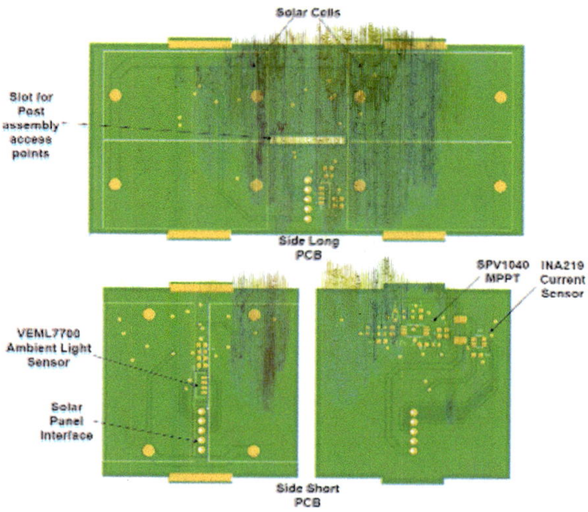

Fig. 10 Long and short side with components

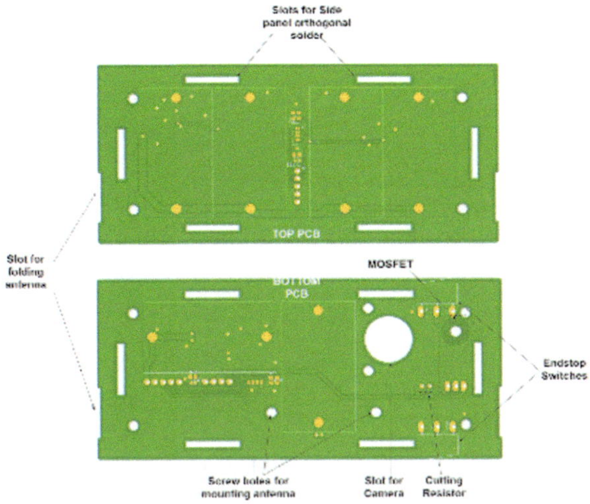

Fig. 11 Top and bottom PCBs with components

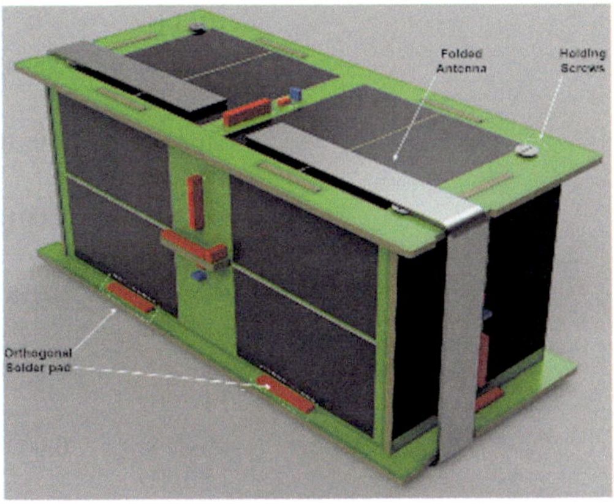

Fig. 12 Assembled view of the proposed design

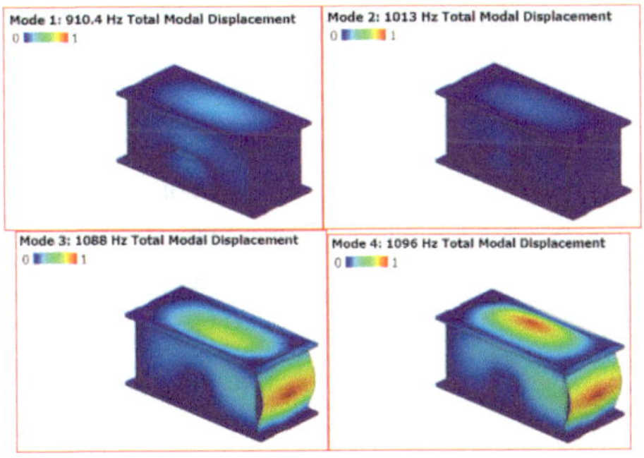

Fig. 13 Modal frequency analysis

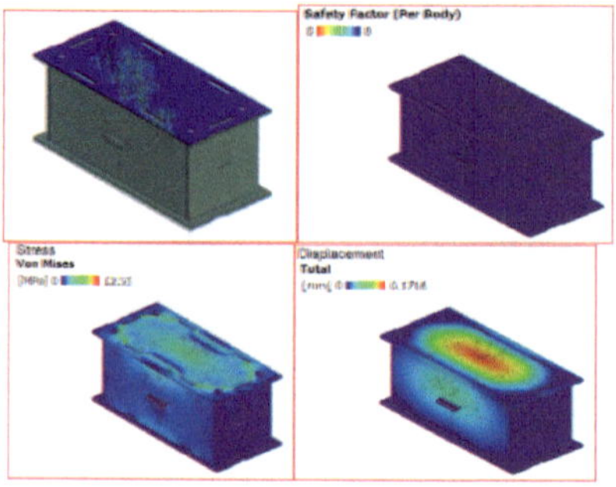

Fig. 14 Static simulation FEA

Table 3 Total modal displacement

Frequency	Participation X	Participation Y	Participation Z
Mode 1: 910.4 Hz	0.0003	0.0004	5.45810014
Mode 2: 1013 Hz	0.0026	0.0002	0.065900001
Mode 3: 1088 Hz	2.89420001	0.0041	3.13329995
Mode 4: 1096 Hz	1.94109995	0.0023	4.49180007
Mode 5: 1220 Hz	0.0018	22.9580998	0.0004
Mode 6: 1268 Hz	0.0181	0.0137	0.71510002
Mode 7: 1470 Hz	0.0378	1.78859998	0.0073
Mode 8: 1510 Hz	0.0028	31.1892003	0.0011

Table 4 Static simulation with 100 N force along $-z$ axis

Name	Minimum	Maximum
Safety factor	4.126	15
Stress		
Von Mises	3.735E−09 MPa	12.53 MPa
First principal	− 6.287 MPa	8.076 MPa
Third principal	− 16.32 MPa	1.106 MPa
Displacement		
Total	0 mm	0.1716 mm
X	− 0.004675 mm	0.004663 mm
Y	− 0.05898 mm	0.05862 mm
Z	− 0.1716 mm	0.03498 mm

5 Conclusion

An open-source pico-satellite can be designed using inexpensive COTS components to serve as technology demonstrator or to host small payloads. Using community backed hardware and platform can significantly reduce the time and efforts required for realizing a mission. Use of COTS hardware in low Earth orbit with innovative designs to tackle challenges due to radiation and temperature can bring down the cost to the range of academia. The observed modal analysis of the structure reveals that the resonating frequency is higher than the recommended PSLV launch environment.

Acknowledgements Team would like to acknowledge K.J.S.I.E.I.T. for funding and supporting the mission. Thanks to Rajeev Senan sir, Kannan S.A sir and Kiran Mohan sir from ISRO for being a continuous guidance source. Special thanks to Tom Walkinshaw, Ben Cartwright, Julian Fernandez, Pierros Papadeas and Joe Laterell from the PQ community for their guidance during design. Thanks to Daniel Estévez for extending gr-satellite reception support for our satellite.

References

1. ISRO. Announcement of opportunity (AO) for Orbital platform, June 2019
2. TU Delft. Alba Orbital, Gauss Srl, The PocketQube Standard Issue 1 June 2018
3. Github repository of, "BeliefSat". https://github.com/NewLeapKjsieit/BeliefSat. Last accessed July 2020
4. IARU. Frequency coordination for spacecraft using the amateur satellite service. Ole Garpestad, IARU, March 2015
5. Arducam. ArduCAM-M-2MP camera shield. ArduCAM-M-2MP, Rev 1.0, Mar 2015
6. IXYS. IXOLARTM high efficiency solarMD\, SM141K06L, Rev. Sept 2018
7. Github repository of Fossasat satellite series, Fossa System
8. STMicroelectronics. High efficiency solar battery charger with embedded MPPT, DS6991, Rev 8-Jan 2020
9. Atmel. Atmega1284p, 8059D–AVR, Nov 2009
10. UKHAS, SSDV specifications
11. Maxim integrated products. Digital temperature sensor and thermal watchdog with 2-wire interface, 19-4385, March 2009
12. Honeywell. 3-axis digital compass IC HMC5883L, 900405, Rev E, Feb 2013
13. InvenSense Inc. MPU-6000 and MPU-6050", PS-MPU-6000A-00, Rev. 3.4., 19 Aug 2013
14. Vishay semiconductors. High accuracy ambient light sensor with I2C interface. 84286, Rev. 1.4, 20 Sept 2019
15. FossaSat-1 Flight data, 9 March 2020
16. Gerhardt DT, Palo SE. Passive magnetic attitude control for CubeSat spacecraft. In: 24th annual AIAA/USU conference on small satellites
17. Product page of NiceRF Lora1268F30

Testing and Qualification of RF MEMS Switch Demonstrator Payload (RFMEMS-TD1) for Small Satellite Platform

M. S. Giridhar, T. K. Pratheek, Shubhajit Biswas, M. M. Nandakishor, Ashwini Jambhalikar, Jiju John, Supriya Gogulapati, Sanjay Mittal, Amit Maji, Deepak Kumar Sharma, Tippayya, S. Ajith Kumar, P. Selvaraj, Kalpana Arvind, A. V. G. Subramanyam, Vamsi Krishna Velidi, Deepankar Roy, A. V. Sucharitha, and V. Senthil Kumar

Abstract This paper reports the design, operating principle, testing and qualification of a compact demonstrator payload designed for evaluating the performance RF micro-electromechanical systems (MEMS) switches in low earth orbit conditions on an experimental small satellite platform. Key aspects calibration of RF level detectors and the approach for fault detection from telemetry data are discussed. The payload has a weight of 800 g, foot print of 100 mm × 110 mm, height 100 mm and power consumption 4.2 W. Telemetry requirements are suitable for typical small satellite missions. The electronics components used are commercial grade which have undergone payload level qualification tests for the intended mission using the proto flight model test philosophy.

Keywords Technology demonstrator · RF MEMS switch · Environmental testing · Data acquisition system · Fault detection

1 Introduction

In recent times, there has been a general trend in the reduction of cost of access to space. Both government run and private space organizations have introduced

---

M. S. Giridhar (✉) · T. K. Pratheek · S. Biswas · M. M. Nandakishor · A. Jambhalikar · J. John · S. Gogulapati · S. Mittal · A. Maji · D. K. Sharma · Tippayya · S. A. Kumar · P. Selvaraj · K. Arvind
Laboratory for Electro-Optic Systems, ISRO, Bangalore 560058, India
e-mail: giri@leos.gov.in

S. Biswas
e-mail: shubhajit_biswas@leos.gov.in

A. V. G. Subramanyam · V. K. Velidi · D. Roy · A. V. Sucharitha · V. S. Kumar
Communications Systems Group, U. R. Rao Satellite Centre, ISRO, Bangalore 560017, India

© The Author(s), under exclusive license to Springer Nature Singapore Pte Ltd. 2023

R.S. Priyadarshini and T. Sundararajan (eds.), *Advances in Small Satellite Technologies*, Lecture Notes in Mechanical Engineering, https://doi.org/10.1007/978-981-19-7474-8_2

programs which allow access to low earth orbits to university laboratories, technical hobby groups and startup companies for experimental purposes. Due to such initiatives, there has been an increase in the development of novel and new technologies for space applications. These technologies have to be validated by proving their performance in actual on board conditions. Small satellites provide a technically suitable and financially viable platform for performance demonstrations of experimental payloads. In this paper, we report the design, operation and testing and evaluation of a compact payload for carrying out demonstration of on-orbit performance of RF MEMS switches fabricated in authors' laboratory. RF switches are critical components in satellite communications, telemetry and payloads modules. RF switches made using MEMS processes are known to be superior than conventional switches [1]. RF MEMS switches have not been used in the Indian Space Programme till date. Prior to induction of these components operational systems, it is proposed to demonstrate their performance in orbital conditions using a compact standalone demonstrator. The payload has all the necessary electronics to operate and monitor the performance of the RF MEMS switches. The payload produces six analog outputs that carry information of operational status of the switches and payload health. The RF section, drive electronics and timing generation module are built from commercial grade components. Payload details can be found in Giridhar et al. [2]. The payload has been designed for flight on PS4 orbital platform (PS4-OP) which is intended to serve as an experimental platform for new technologies. However, the general design and interfaces of this payload make it a suitable candidate for any small satellite project. The details of environmental tests carried out on this payload to qualify it for experimental flight with a mission life of about 6 months are described in following sections. A similar RF MEMS switch demonstrator payload has been flown in 2014, but results are not yet seen in the literature [3, 4]. RF MEMS switches were also launched on sounding rockets to study the effects of launch conditions on these devices but these only high altitude flights and not orbital platforms [5].

2 RF MEMS Switch

The RF MEMS switches described in this work are of the ohmic contact type. They can be described as chip level transmission lines on a glass substrate with a gap in the signal carrying line. A micromachined metal contact is suspended above the gap. The contact is part of a larger suspended structure called the actuator. The actuator is suspended by means of compliant micromachined flexures. Metal electrodes are positioned below the actuator on the glass substrate. When a DC voltage (exceeding a certain threshold) is applied across the actuator and the electrodes, the actuator snaps down on the transmission line bridging the gap with the contact. This action takes the switch into the ON state where the input and output ports get connected. When the applied voltage is removed, the restoring force in the flexures pull the actuator away from the transmission lines taking the switch to its OFF position. The switches described in this work are fabricated in the authors' laboratory. Complete

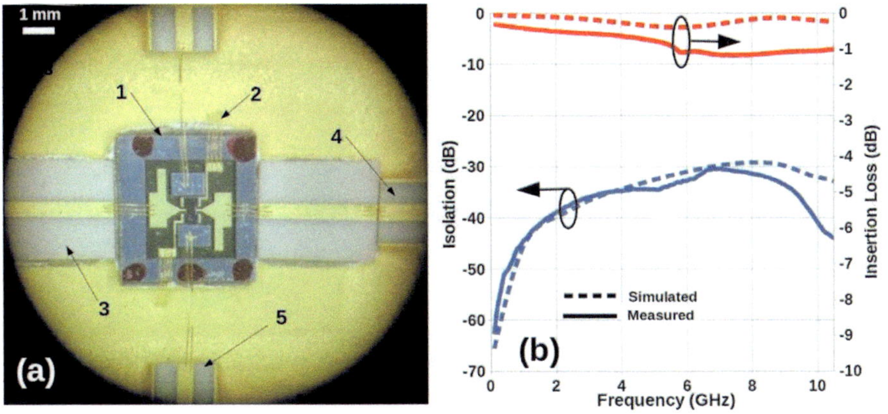

Fig. 1 **a** Photograph of the RF MEMS switch assembled in the RF package; 1-MEMS switch, 2-ground wires, 3-link card, 4-package RF terminal, 5-package DC terminal **b** RF characteristics of the packaged RF MEMS switch in the ON and OFF states. The switch has an insertion loss of about − 1.0 dB and isolation better than − 40 dB at 10 GHz

details of the fabrication process are described in Giridhar et al. [6] and references therein. The fabrication is based on the silicon-on-glass architecture, adapted from the work of Kim Yong-Kweon of Seoul National University [7]. In this fabrication process, deep reactive ion etching (DRIE) is used to fabricate the mechanical micro-structure in single crystal silicon. The transmission line and electrodes are fabricated on glass substrate. The silicon and glass substrates are then bonded by anodic bonding and processed further to realize the final structure. The fabricated switches are screened, tested and packaged in hermetic RF packages. These packages can be used in microwave circuits where the control and routing of RF signals are required. A photograph of the RF MEMS chip assembled inside the package and measured RF characteristics are given in Fig. 1.

3 Payload Description

The block schematic of the payload is given in Fig. 2a. The main modules of the payload are individually described in this section.

3.1 RF Module

The RF module comprises of the RF MEMS switches, RF source, power divider and RF level detector. The signal source is a voltage controlled oscillator whose output is fixed to produce a signal at 363 MHz at a power level of + 11 dBm (PEIV31009). This

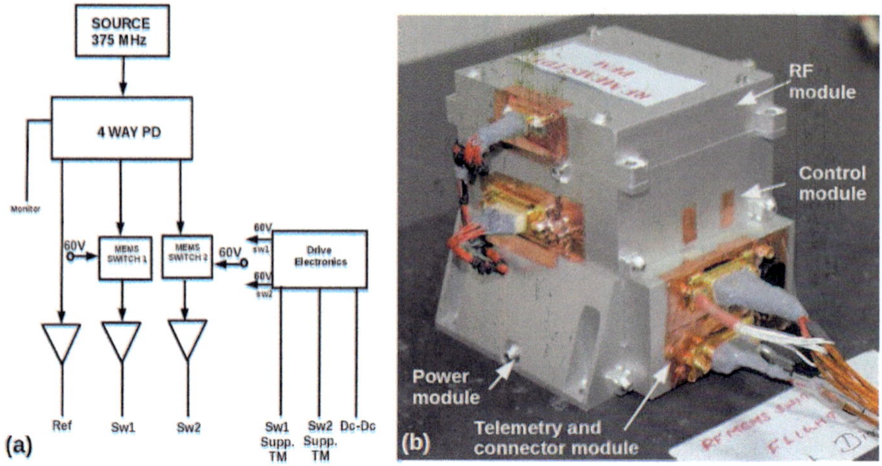

Fig. 2 **a** Block diagram of the payload **b** photograph of RFMEMS-TD1

test signal is fed to a four way power divider. Two outputs of the 4 way power divider (4 way PD) are connected to individual RF MEMS switches. The outputs of the RF MEMS switches are fed to two independent level detectors. The RF level detectors are log-linear detectors (LT5537). The output of the detector is a DC voltage that is linearly proportional to the input RF power (in dB units). The third output is fed directly to an RF level detector. This signal serves as a reference signal. This signal is used to monitor the source power. The fourth output is fed to an SMA connector. This output is used to monitor the source on ground and will be terminated with a 50 Ω termination before flight.

3.2 Drive and Control Module

Electrostatic RF MEMS switches require high voltages for operation. The switches described in this work require an actuation voltage of 60 V. A custom voltage booster circuit has been developed to generate the high voltage from standard sources. The drive electronics not only produces the actuation voltage but also latches the voltage to the appropriate switch based on pulse commands. High voltage MOSFETs are used to the deliver the actuation voltage to the switches. The circuit also has protective features that ensure that the applied voltage does not exceed 60 V even if the input voltage becomes higher than expected. The triggering pulses for the drive module are produced by an FPGA programmed to alternately turn the MEMS switches ON and OFF with a period of 10 s in a complementary manner.

3.3 Power Module

Electrical power to the payload is provided by a DC–DC converter capable of running from a raw bus in the range of 28–42 V. The converter provides a stable output of ± 15 V for payload operation. An EMI/EMC filter is used on the input side of the converter to minimize conducted emissions into the raw bus.

3.4 Telemetry Outputs

The payload has the capability to provide both digital and analog telemetry. The FPGA (Smart Fusion) used in the control module has built-in analog to digital converters that can digitize the level detector outputs and transmit the data on an SPI bus. This serial data is fed to a level shiftier to produce signals compliant with the RS485 standard. RS485 is a differential communication technique which is highly immune to noise pick up. The present payload is configured for analog telemetry since the spacecraft followed a custom digital format. The analog signals are fed to the spacecraft data processing system for digitization and transmission to ground via the spacecraft telemetry system.

The circuit boards of the various modules are assembled in a stacked manner in a lightweight aluminum housing as shown in the photograph in Fig. 2b. The modular design allows the RF module to detached from the rest of the payload allowing to be independently tested or swapped with another module if required.

4 Bench Model: Source Characterization and RF Level Detector Calibration

A critical step in the development of the payload was the testing and calibration of the RF level detectors. It was important to establish the suitability of the RF level detector in terms bandwidth and input power ranges. The RF level detector circuit used in the payload is based on the evaluation board supplied by the manufacturer.

A bench model of the RF module was made using connectorized versions of the same components that were proposed for the proto flight model. Using the bench model, the RF components were tested for their compatibility with the RF MEMS switches. The bench model was used for calibration of the RF level detector using RF synthesizers. The RF source was characterized with respect to harmonics, tunability and output power. Results of the level detector characterization are shown in Fig. 3. It can be seen from Fig. 3a that in the ON and OFF states of the switch results in a swing of 2.1 V in the detector output over a wide range of input powers from − 15 to + 10 dBm. The RF spectrum at the output of the switch for + 5 dBm input power is shown in Fig. 3b. The VCO, which is the RF signal source, is tested for its output

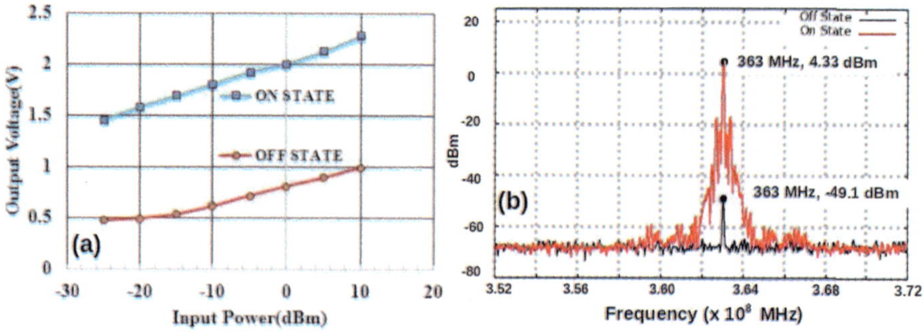

Fig. 3 **a** RF level detector response to input power sweep, at 360 MHz, in ON and OFF states of the switch. The dashed line indicates the operating RF power of the switches in the payload **b** RF power spectrum at the output of the switch in ON and OFF

frequency and power with varying control voltage, Fig. 4a. In the payload, the VCO control voltage was fixed to 15 V, directly available from the DC–DC converter, resulting in an output frequency of 363 MHz. The spectrum of harmonics from the source is shown in Fig. 4b. It can be seen that in comparison with the fundamental the contribution to output power from harmonics is negligible. Photographs of the payload in various stages of assembly are shown in Fig. 5

5 Payload Operation and Fault Detection

The payload is designed for autonomous operation with no intervention from ground. The switches start cycling in a complementary manner with a preset period as soon as power is made available to the payload. The status of the switches can be deduced from the values received in the telemetry data as shown in Table 1. The number of cycles can be estimated from the time elapsed after "power up" of the payload.

6 Environmental Tests for Low Earth Orbits

The objective of the environmental tests is to simulate the conditions that will be experienced by the package during storage, launch and orbital phases of the mission life of the satellite. Typically, two identical models of the payload are built. They are designated as qualification model and flight model. The qualification model is subjected higher levels of environmental stresses and any anomalies during these tests are corrected and tests repeated to ensure that corrective actions are adequate. The same corrections are carried out in flight model, and it is subjected to same tests at lower stress levels but for longer duration. The flight model, as the name indicates, is the one that is integrated in the spacecraft. However, in some instances, due to

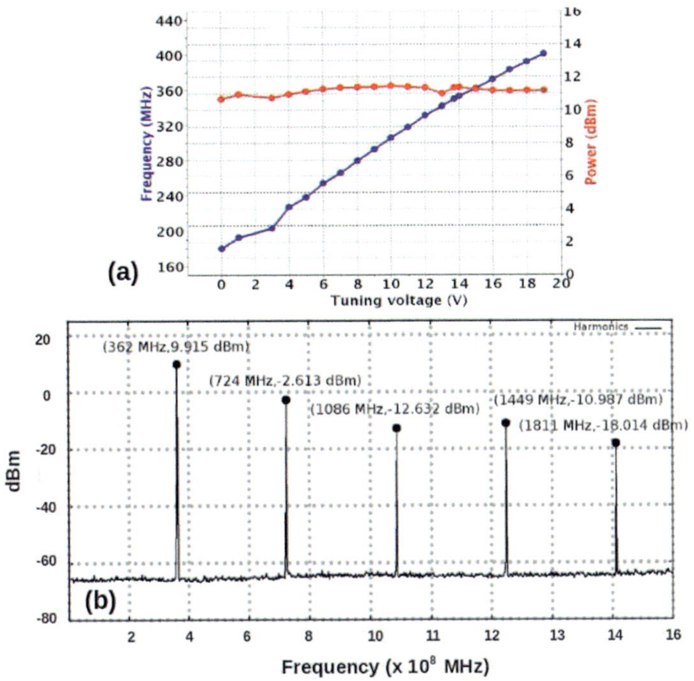

Fig. 4 **a** Output frequency and power measurements for VCO used as RF source. **b** Measurement of harmonics of the source

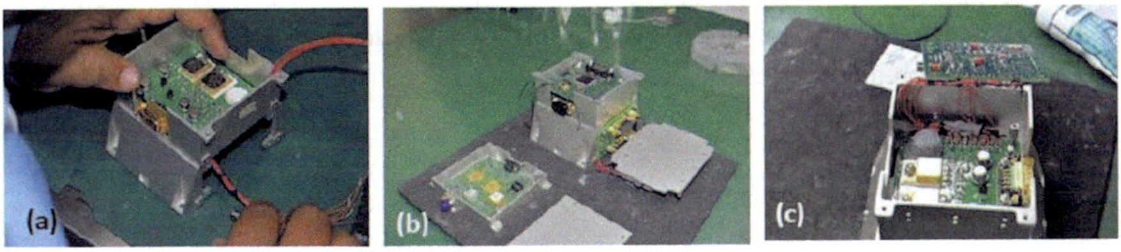

Fig. 5 Photographs of RFMEMS-TD1 at various stages of assembly **a** power module integration **b** RF module assembly **c** FPGA card assembly into the housing

resource constraints only one model may be built. This is particularly true in case of experimental payloads for small satellites. In such a situation, the single model is called a proto flight model (PFM). PFMs are subjected to qualification levels for acceptance duration. Certain highly stressful tests such as mechanical shock are not conducted on the PFM. The payload described in this work RFMEMS-TD1 is a PFM.

The following paragraphs describe the environmental tests carried out on the RFMEMS-TD1 PFM payload.

Table 1 Status of the RF MEMS switch indicated by telemetry values

Telemetry channels				MEMS switch status
Switch output (V)	Actuation status (V)	Reference (V)	DC–DC status (V)	
2.05–1.09	5.0	2.04	5.0	Normal ON state
1.09–0.8	5.0	2.04	5.0	Degraded ON state
> 0.8	0.0	2.04	5.0	Normal OFF state
> 0.8	5.0	2.04	5.0	Failed/degraded OFF state
1.09–2.05	0.0	2.04	5.0	Failed in ON state
XX	XX	< 2.04	5.0	Source power degradation

6.1 Thermal Cycling

In this test, the payload is subjected to hot and cold temperatures at ambient atmospheric conditions. During thermal cycling, all components undergo periodic thermal expansion and contraction cycles. This effect will precipitate assembly problems such a dry solder on the circuit board. This is an "active" test. The payload is powered for short durations during the hot and cold cycles to check for deviation from normal operation and effects of internal heat dissipation. Thermal cycling is preceded by hot soak (60 °C for 24 h) to ensure the degassing of volatile assembly materials used in the payload such as epoxies and sealants.

6.2 Vibration Test

This test subjects the package to mechanical vibrations as experienced during integration and launch. The *Z*-axis is assumed to be the direction of launch. Vibration test precipitates weaknesses in mechanical assembly. In addition to workmanship issues, this test also verifies that all mechanical resonances are sufficiently far away from the vibrations experienced during launch. Vibrations are imparted in two ways, sinusoidal sweep and random vibrations, sequentially along the three Cartesian axes of the payload. Accelerometers are attached to the payload at vulnerable locations record displacement and accelerations during the test. For the RFMEMS-TD1, vibration testing is done in a passive mode, i.e., the packaged RF MEMS switch is powered OFF during vibration. This is because in the mission plan, package will be powered OFF during launch and orbit insertions phases.

6.3 Thermo-Vacuum Testing

In thermo-vacuum testing, the payload is subjected to thermal cycling under vacuum conditions. This environment mimics the orbit conditions that will be experienced by the payload. The temperature excursions occur during eclipse and sunlit conditions experienced by the payload. The main difference between thermal cycling and thermo-vacuum is that in thermo-vacuum, convective heat losses from and within the payload are nearly absent. In this situation, any hot spots generated during the operation of the package, will be localized. If any locations in the payload are suspected to develop hot spots, temperature sensors are placed in their vicinity for monitoring in addition to sensors for general package temperature monitoring. For the RFMEMS-TD1 payload, additional sensors were placed near the DC-DC convertor since it has the highest dissipation in the payload.

6.4 Electromagnetic Compatibility (EMC) Tests

All packages and sub-systems that are part of the satellite have to undergo electromagnetic interference (EMI) and susceptibility tests. Emissions are tested in both conductive and radiative modes over a wide band of frequencies. The objective of this test is to ensure that the package does not cause harmful interference to other satellite sub-systems. The second objective of the test is to determine the susceptibility of the package to external electromagnetic radiation impinging on it. This test is done in active mode where the output is monitored when the payload is subjected various electromagnetic radiative fields. Conductive tests are performed by injecting noise into the raw bus power lines.

Photographs of the payload undergoing various tests are given in Fig. 6. The important test parameters for the tests outlined above are given in Table 2.

7 Payload Data Acquisition System

As mentioned previously, the RFMEMS-TD1 payload produces six analog output channels that require to be monitored on ground using spacecraft telemetry. For ground testing, a compact six channel data acquisition and logging system was developed. The six analog outputs from the payload are fed to six ADC channels of a proto typing board (Arduino Uno). The digitized data is plotted and logged by a single board computer (Raspberry PI model 3) connected to the proto typing board. The computer triggers the proto typing board to execute an acquisition cycle after completing plotting and logging tasks. The RF MEMS switch outputs are tapped using an online monitor and fed to an oscilloscope. The oscilloscope is required for observing noise levels in the output signals which is not possible for the digitizer

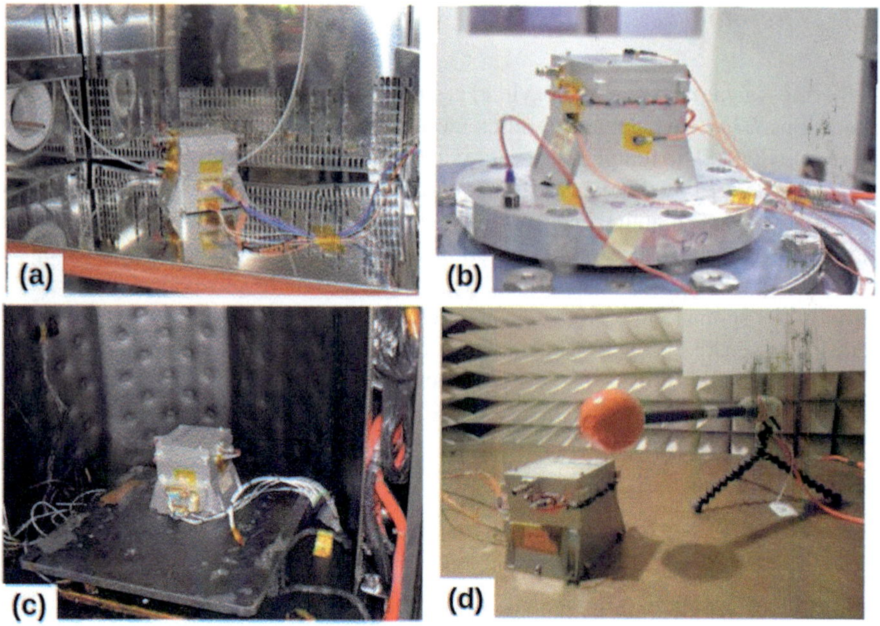

Fig. 6 **a** Payload loaded in the thermal cycling **b** Mounted on the shaker table for *z*-axis vibration testing **c** Inside the thermo-vacuum chamber **d** Payload undergoing EMI/EMC testing, one of the antennae for emission tests is shown in the photograph

Table 2 Key parameters of the qualification tests for RFMEMS-TD1

Sl. No.	Test name	Test parameters
1	Thermal cycling	+ 5 °C to + 70 °C, 2 cycles, dwell time 2 h + 40 °C 2 h
2	Vibration (sine)	*X*, *Y*, 10–100 Hz 7.73 g max, 1 oct/min, duration 1 min
		Z 10–199 Hz 9.99 g max, 1 oct/min, duration 1 min
3	Vibration (random)	*X*, *Y*, *Z* 20–2000 Hz max amplitude 0.034 mm, g 13.5, duration 1 min
4	Thermo-vacuum cycling	+ 5 to + 70 °C dwell time 90 min, chamber pressure 2 × $10e^{-6}$ mbar
5	EMI/EMC tests	Radiative susceptibility test (RS03) 30 MHz–2 GHz, 5 V/m field strength
		Conductive susceptibility tests CS01: 30–150 kHz CS02: 15–50 kHz, 1 V rms noise
		Radiated emissions tests RE02: 14 kHz–18 GHz Conductive emission test CE03: 15 kHz–50 MHz

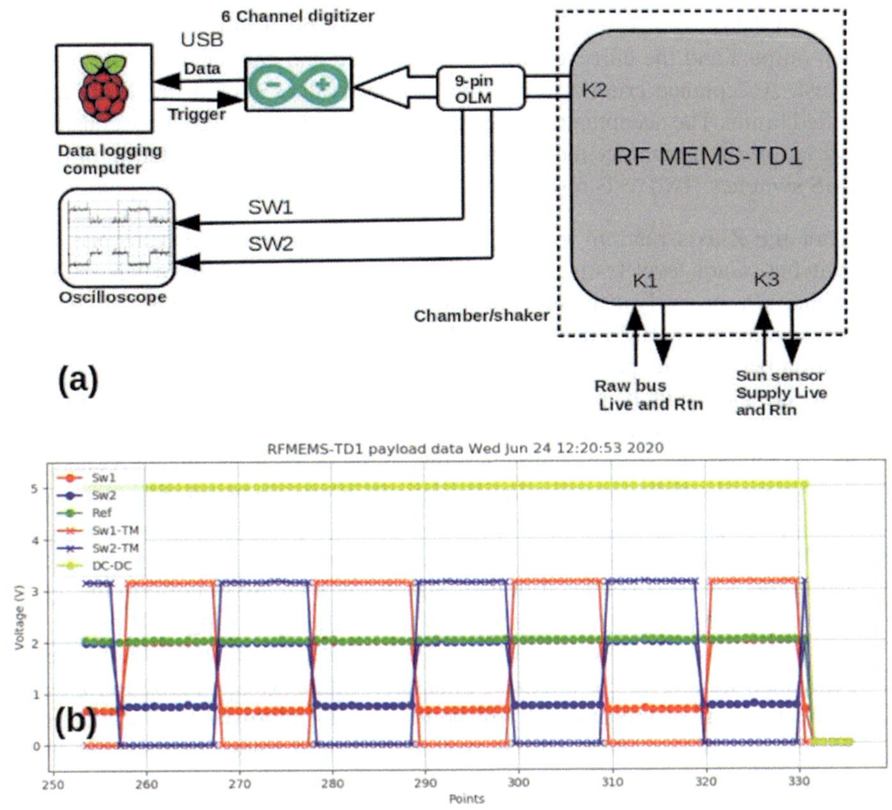

Fig. 7 **a** Schematic of the payload acquisition set up. **b** Typical live plot of payload operation produced by the data acquisition system

since it has low sampling rate. The switches are cycled ON and OFF with period of ten seconds. The data acquisition systems are compact and built from open source software and hardware elements. A GUI using Python was developed for ease of operation. The schematic of the data acquisition system and a plot of the live data obtained as shown in Fig. 7.

8 Results and Discussions

The main parameters of interest in the payload data are the outputs of the two RF MEMS switches. The outputs are DC voltages produced by the RF level detectors. The RF level detector produces a voltage proportional to the RF power it receives in dB units. Both ON and OFF states of the switches are monitored to ascertain the health of the switches. In addition to the switch outputs, the current drawn at

28 V from the raw bus is also monitored as a sign of the health of the payload. The switch outputs and the current drawn are measured and compared for the pre- and post-test. Acceptance criteria is that the post-test values should lie within certain specified limits. The acceptance ranges given in Table 3 are arrived at by considering the variations in laboratory measurements of insertion loss and isolation of the RF MEMS switches. Two tests resulted in observations in the payload outputs.

- After the *Z*-axis random vibrations, no output was observed from one of the switches. Card level testing revealed that for the RF MEMS switch, actuation voltage was properly available. Further, testing indicated that the RF ribbon bond between the RF MEMS package and PCB may have got detached due to relative motion between the package and PCB induced by the vibrations. It was decided to secure the RF PCB in the vicinity of RF MEMS switches. The vibration tests were repeated with accelerometers attached close to the MEMS switches in both the unsecured and secured configurations of the PCB. A significant reduction in the magnitude of the vibration levels of the PCB was observed after this rework. Subsequently, the package cleared the *Z*-axis random vibration test.
- In EMI/EMC tests, susceptibility was found at some at about 2 MHz (conducted) and in the 3–4 GHz (radiated) range. Significant increase in noise in the analog outputs of the payload was observed in these frequency ranges. However, this interference did not affect overall operation of the payload and did not have any permanent effects on the payload. This observation will be borne in mind and future versions of the payload will be made with improved housing to reduce RF leakage. Better filters will be used in the power lines to reduce conducted susceptibility.

The post-test values and the acceptance values for switch outputs in the ON and OFF states are given in Table 3.

Table 3 Post-test performance results

Heading level	Switch state	Measured value (V)	Acceptance range (V)
Switch 1	ON	2.04	1.09–2.05
	OFF	0.67	0.5–0.8
Switch 2	ON	2.03	1.09–2.05
	OFF	0.67	0.5–0.8
Reference	–	2.06	1.09–2.05
Raw bus	28 V	150 mA	150 ± 5% mA

9 Conclusion

A compact payload for on-orbit demonstration of operation of RF MEMS switches has been built with commercial components and qualified for typical small satellite missions. A simple analog data acquisition and logging system has been developed for checkout and ground test activities. Test details are discussed, and post-test performance results are reported. The payload is awaiting flight opportunity on a small satellite platform.

Acknowledgements The authors are thankful to C.V.N. Rao and Shailendra Singh of Space Applications Center, ISRO for chip and package level RF design and simulations of MEMS switch. The authors gratefully acknowledge fabrication support by G. Sai Sarvanan and packaging support from Sangam Bhalke of Gallium Arsenide Technology Enabling Center, Hyderabad. They are very thankful to B.H.M. Darukesha of U.R. Rao Satellite Center for valuable help in the assembly requirements of the RF module.

References

1. Rebeiz GM (2003) RF MEMS: theory, design, and technology, 1st edn. Wiley, New Jersey
2. Giridhar MS, Joju R, John J, Jambhalikar A, Supriya G, Pratheek TK, Bhuvan MB, Mittal S, Viswanathan M, Subramanyam AVG, Velidi VK, Sucharitha AV, Roy D, Senthil Kumar V (2018) Development of demonstrator payload for evaluating on-orbit performance of RFMEMS switches. In: Proceedings of ISSS national conference on MEMS, smart materials, structures and systems, October
3. Cazaux JL, Vendier O, Mesthé L, Courtade F (2013) MEMO flight experiment delivered! http://www.microwave-rf.com/docs/12h15-THALES-ALENIA-SPACE.pdf
4. Vendier O, Mesthé L, Lavalette V, Cazaux JL, Villemazet JF, Reig B, Blondy P, Courtade F, Pressecq F (2014) Validation of the MEMO experiment at satellite level: RF MEMS GEO flight demonstration. In: 9th ESA round table on micro and nano technologies for space applications, Lausanne, Switzerland, 10–13 June 2014
5. O'Mahony C, Olszewski O, Hill R, Houlihan R, Ryan C, Rodgers K, Kelleher C, Duane R, Hill M (2014) Reliability assessment of MEMS switches for space applications: laboratory and launch testing. J Micromech Microeng 24(12):125009
6. Giridhar MS, Jambhalikar A, Jiju John R, Islam AB, Nagendra CL, George Thachil MP, Srikanth SS, Darukesha BHM, Bollu S (2013) Design, fabrication, testing and packaging of a silicon micromachined radio frequency microelectromechanical series (RF MEMS) switch. Sadhana 38:297–316
7. Jong-Man K, Jae-Hyoung P, Chang-Wook B, Yong-Kweon K (2004) The SiOG-based single crystalline silicon (SCS) RF MEMS switch with uniform characteristics. J Microelectromech Syst 13:1036

Student Small Satellite Design and Development in India: A Review

Sini V. Pillai, R. S. Priyadarsini, and Ranjith S. Kumar

Abstract Student-built space missions offer prospects to commence small space missions with small financial budgets. The significant outcome along with technical intrinsic worth is the cross-disciplinary training for students that endow a future career in the space industry. The experience in learning and working with hands-on, project-based education proves effective for talent management among the student community. Programs such as the CubeSat standard have dramatically changed spacecraft engineering education, fostering aerospace education resulting in designing and building small satellites. The paper explores the proliferation of the status of the University/College supported student-built space missions over the last years and the various levels of applications of small satellite platforms in India. Moreover, how the student satellite projects translated into useful small satellite commercial missions is also discussed.

Keywords Satellites · Miniaturization · Student satellite projects

1 Introduction

In the past two decades, the ease of access to efficient and reasonable Commercial Off-The-Shelf (COTS) microsystems paved way for revolution into the field of small

S. V. Pillai (✉)
CET-SoM, College of Engineering Trivandrum, Trivandrum, Kerala 695016, India
e-mail: sini.mba@cet.ac.in

R. S. Priyadarsini
Department of Civil Engineering, College of Engineering Trivandrum, Trivandrum, Kerala 695016, India
e-mail: rspriyadarsini@cet.ac.in

R. S. Kumar
CET-Centre for Interdisciplinary Research, College of Engineering Trivandrum, Trivandrum 695016, India
e-mail: ranjith@cet.ac.in

© The Author(s), under exclusive license to Springer Nature Singapore Pte Ltd. 2023

R.S. Priyadarshini and T. Sundararajan (eds.), *Advances in Small Satellite Technologies*, Lecture Notes in Mechanical Engineering, https://doi.org/10.1007/978-981-19-7474-8_3

satellite systems worldwide [1]. Typical classification of this low-mass artificial satellite systems is as follows, small-satellites (< 1000 kg), microsatellites (10–100 kg), nanosatellites (1 – 10 kg), and picosatellites (0.1 – 0.99 kg). Many successful attempts have been performed by different educational institutions around the globe to realize nanosatellites especially. Engineering students of Aalborg University with more than 150 students from different master specializations have been building student satellites since 2001. The first satellite AAU CubeSat was launched in 2003; second one a pan-European student satellite, SSETI Express, was launched in 2005; and the third satellite, a pico-size satellite named AAUSAT-II, was launched from Satish Dhawan Space Centre, India, during 2008 [2]. The CubeSat Program which started in 1999 at Stanford University has turn out to be a world-wide program comprising over 70 universities along with government and private organizations, commenced with an educational perspective aiding to develop very low cost, low weight satellite within two years to train students in industry's multidisciplinary environment [3].

The space industry has shifted their focus from producing large sophisticated spacecrafts toward advancements in developing small satellites with CubeSat standard. The CubeSats at first were considered as low-cost educational demonstration satellite platforms, later advanced CubeSat missions were developed which shifted from sole educational purposes to development of low cost, high value, and real-time scientific missions fly on a capable spacecraft [4]. The CubeSat standard focused on reduction of development cost, and the first CubeSat was launched in 2003, enabling hundreds of missions by its standardization [5]. The term small satellites refers to those with low size and mass, usually under 500 kg. The huge cost of launch vehicles and those associated with the manufacture can be reduced by miniaturizing the satellites. Small satellites permit cheaper design, facilitating mass production, and enable missions like in-orbit check and constellations of low data rate communications, which the larger satellite possibly will not carry out. The main technical confront involved in the construction of small satellites includes lack of sufficient power storage or room for a propulsion system.

The space program in India is considered to be the largest national space program in the world witnessing the transition from pure research direction to commercial purposes since the 1990s. In India, small satellites are developed by Indian Space Research Organization (ISRO) to carry out experiments in space. Small satellites provide enormous research opportunities for the development of a country by tapping innovative young talent. ISRO holds various student-developed satellites with variant missions for space applications by constantly encouraging capacity building among the student community to prepare for future space technology applications of designing, developing, fabricating, and testing the space technology systems and subsystems. Various Universities and academic institutions started introducing space science and space technology applications in their educational curriculum consequently evoking interest among students to design satellites for various space applications. The purpose is to train students in design, integration, and operation to give direct control over the various steps to take up the development aspects forward in the program. ISRO managed to coordinate and streamline the space technology activities of students from various universities in the form of small satellites as part of Indian remote sensing program in effect which facilitated ISRO to build and launch various small satellites collaborating with universities within and outside the country. KALAMSAT-2, NIUSAT, PRATHAM, PISAT, SATHYABAMASAT, SWAYAM, SRMSAT, JUGNU, STUDSAT, and ANUSAT are the various student small satellites successfully launched till date by ISRO.

Various areas supported by ISRO in small satellite missions for academia are in the areas of design and development of the payload, provides the design for a standard nanosatellite bus fabricated by the institute, and provides mission-critical components/materials such as solar panels, batteries, multi-layer insulation, thermistor, etc. and finally conduct testing of the small satellite developed by the institute under the technical guidance of ISRO. In addition, ISRO has launched a program called Young Scientist Program in which three students from each state as well as from the Union territories will be selected for a one-month program in ISRO selected centers based on their performance in science projects. The students will be trained by senior space scientists, and opportunities will be provided to design small satellites subjected to launch based on quality by making use of the R&D facility (6). In this review, the time evolution of different student satellite programs is detailed along with their outcomes. Moreover, the grooming of small satellite commercial applications as an outcome of ISROs different initiatives including student satellite programs also was discussed.

2 Student Satellites in India—An Overview

India also falled in line with other developed countries so as to encourage students to launch their own satellite systems. Anna University is the first Indian University to develop a satellite named Anna University Satellite (ANUSAT) which demonstrated the technologies associated with message storing and forward operations under the overall leadership of ISRO and was launched in 2009. This success indeed motivated other Universities to develop small satellites for space applications. A team of students belonging to seven engineering colleges from Karnataka and Andhra Pradesh developed the first picosatellite weighing lower than 1 kg, named as Student Satellite STUDSAT [6]. The mission of the satellite includes promoting space technology in educational establishments, support research activities, and progress in miniaturized satellites by creating communication links between the ground station and the satellite to capture a range of images of the earth with resolution 90 m and finally to broadcast payload to the earth station. The STUDSAT was launched with ISRO's PSLV C-15 on July 12, 2010 [6]. StudSat-1 is the first picosatellite of experimental nature with a mission life of six month and is the smallest satellite launched by India. The students were able to get hands-on experience with the design and fabrication at a minimum cost. STUDSAT-2 is India's First Twin Satellite Mission consisting of two Nano Satellites, STUDSAT-2A and STUDSAT-2B with dimension of 300 × 300 × 150 mm, weight 6 kg and 5.5 kg for Inter-Satellite Communication [7]. Each satellite has two payloads with complementary metal oxide semiconductor (CMOS) camera, resolution of 80 m per pixel with inter-satellite link (ISL) arrangement [8]. The nanosatellites are developed using reliable and robust real-time operating system (RTOS) that includes task management with priority assignment and scheduling functionality using ARM CORTEX-M4 platform [8]. The separation mechanism system is interconnected between STUDSAT-2A and STUDSAT-2B and after ejection from the PSLV, the satellite is expected to stay composite for an estimated duration of three orbits prior to separation due to stability constraints. It is of greatest importance to establish an unfailing separation mechanism to withstand the vibrations of the launch vehicle with the space environment.

A joint involvement of students from Indian and Russian Universities developed a mini satellite of lift mass 92 kg named as Youthsat which is considered as the second accomplishment in the Indian Mini Satellite series. Youthsat investigates the association between solar variability and the changes in thermosphere-ionosphere (2). A satellite with the mission of collecting data on greenhouse gases, Sathyabamasat, was designed and developed by the students of Sathyabama University, Chennai. NIUSAT is the satellite designed and developed by Noorul Islam University Tamilnadu with payload of a four spectral band wide field Sensor [9]. Rashtreeya Vidyalaya Satellite (RVSAT-1) was developed by the students of R V College of Engineering, Bengaluru, and is the first nanosatellite from India to carry a mass of microbes to space [20].

The students and faculty of SRM University developed a nanosatellite weighing 10.9 kg named as SRMSat, which can tackle the predicament of global warming in the atmosphere. Vikramsat, satellite for biological experiments in space developed by Space Kidz, imaging satellite Microsat-R for DRDO, and a small communication satellite Kamalsat with life span of two months weighing 1.2 kg which is considered as the world's lightest and first ever 3D-printed satellite developed by students and Space Kidz India, are being launched by ISRO as student made satellite in the space program 2019 (6). Kalamsat is the world's lightest and smallest satellite named after the former President Dr A. P. J. Abdul Kalam (6). A new-fangled variant of India's Polar Satellite Launch Vehicle (PSLV) rocket carried the satellites, and ISROs launch mission PSLV-C44 was the first in the world to employ rocket viewing it as a platform meant for experiments in space. ISRO is also launching another communication satellite GSAT 31 by Ariane space rocket (3). The students of the Indian Institute of Technology, Kanpur, under the leadership and control of ISRO developed a nanosatellite Jugnu of 3 kg mass which can image the Earth in the near infrared region and carry out test using image processing algorithms (2). PISAT was designed by a consortium of Indian colleges led by PES University along with the support of ISRO. PISAT with payload of 166 gm, CMOS-based camera covering 185 km × 135 km area with 90 m spatial resolution was developed to provide hands-on environment in all aspects of satellite building for students [9].

IIT Bombay has taken up a Student Satellite Program (IITBSSP) to develop a Center of Excellence in Satellite and Space Technology with an interdisciplinary team of students from the institute. The first satellite under the space program is completely planned, designed, built, and tested by the students and faculty of IIT Bombay with the assistance of ISRO and named as Pratham [10], identified as the first step towards infinity with the mission of GPS correction. The Student Satellite Team of IIT Bombay aspires in developing the lightest actively controlled satellite in India [11]. Miniaturization play a fundamental role in developing low-cost small satellite missions for space exploration projects [6]. The team researches an approach towards mass minimization in the design of satellites maintaining structural integrity of the satellite in the launch and in-orbit environment [11]. Pratham confronted limited sensing with effective computational and communication capability, motivating the call for autonomous and computationally efficient algorithms [12]. Pratham was launched onboard PSLV-C-35 in 2016 with a mission life of 4 months. Even though the satellite was operational, the mission was not successful and doing well (12). The team could identify the relevance of quality by executing a post-launch analysis; further, the student satellite team worked on instituting quality assurance and good management practices throughout the development. Pratham facilitated teaching students withinside the area of Satellite and Space Technology specifically in the procedure of constructing a satellite for calculating total electron count (TEC) of the ionosphere. Ground station layout used by Pratham with the dimension of TEC of 99.9 accuracies done by using the precept of Faraday rotation. The technique helps in lowering the price of the ground station in addition to the onboard complexity. The satellite will permit the technology of TEC maps of India and France with the assistance of a community of ground stations installation at 11 Indian universities and the Institut de Physique du Globe de Paris (IPGP), France. Pratham can transmit onboard records when it passes over India [13]. Pratham has furnished valuable knowledge of possibility to a massive variety of college students of IITB. The social intention of Pratham represents the notion of the team that knowledge received withinside the area of satellite technology is an essential constituent of mission targets and challenge success in terms of student satellites [14]. The second satellite Advitiy is designed based on the experience gained from the first launch of student satellite Pratham with the mission to transmit stored and uplinked images through SSTV by designing an antenna deployment system (5). The Ham Radio Club which is part of the IITBSSP aims to maintain a functional ground station promoting amateur radio activities in the institute and also organizes an annual Workshop for engineering colleges around the country to support setting up their own ground station. The IITB team is also working on a lunar mission called The Great Lunar Expedition for Everyone (GLEE) which is an initiative by the University of Colorado Boulder to deploy LunaSats (5).

Students from interdisciplinary areas engage in the research and development process of small satellites as the technical background is a requisite for the organization and management of resources for satellite project implementations. The COEP Satellite Initiative (CSAT) is a student satellite program by the students of College Of Engineering Pune, India, and developed a 1U picosatellite named Swayam, the first project under this initiative [15]. The team consisted of students working in functional subsystems which focus on developing the functional aspects of the satellite and systems engineering groups looking into the cohesive planning and integration of subsystems, and the quality assurance and control team investigating workmanship and verifying hardware and software performance for reliability aspects.

3 Small Satellites Development by the Industry

The Internet has revolutionized and changed the outlook of well-being. Similarly, small satellites are expected to take up endless possibilities from the practical applications with the unique ability to cover any place on the earth's surface. Micro and nanosatellites are designed and developed for space missions devoid of spending an enormous sum of money. From 1994 to 2017, there have been 344 university-class spacecraft launched from 166 educational institutions of 47 countries worldwide [16]. Nanosatellites are incredible to many educational institutions to get involved in space missions in terms of affordability. Picosatellites weighing fewer than 1 kg are used for commercial applications. Small satellites can serve military, humanitarian, and civil purposes, especially in crisis situations when the space industry gets implanted with industrialization. The PalmSAT is a picosatellite program at the Surrey Space Centre, University of Surrey, UK developed small satellites for employing in the scientific, commercial and military market [17]. The development in microelectronics along with electric micro propulsion and other small satellite subsystems, enabled the development of a multitude of interplanetary high-power Cubesat and nanosatellite missions [18]. New technologies and advances in small satellites including computational technologies provide operational services creating opportunities for industry players and startups in the areas of space technology. The CubeSat programs along with the student-developed small satellite mission programs have spurred the growth of various entrepreneurial-minded businesses targeting niche markets in India. Many new companies popped up along with the existing engineering companies developing specific components/subsystems for small satellite communities.

The startups which are considered as the future potential of the space sector are discussed here. Planet Labs, an American company is considered as one of the most mature startups in the small-satellite space having 149 satellites in orbit enabling data collection to build and sell data sets to stakeholders across many industries. The core mission is to capture imagery of the whole surface of the earth each and every day. Spire is a startup company considered as the leader in using satellite data to notify weather predictions using data analytics supports the maritime industry and aviation sector. Spaceflight company puts forward small-satellite sector's pricing for various commercial launches and offers end-to-end management services ensuring completion of satellite mission on time and on budget. Spaceflight launched 112 satellites till date from 32 countries and is still counting. Satellogic offers satellite data services with high-resolution geo-intelligence in agriculture, oil pipeline monitoring, disaster response/preparedness. Satellogic has six satellites in the orbit and is set up to reach 300 toward the end of the decade. Vector space systems provides affordable space access to technology companies to send innovations and applications to space. Startup Terran Orbital specializes in nanosatellite design and its development along with manufacturing, testing, and further launching, with a special focus on using nanosats and satellite-secured data for military and defense intelligence applications. Nanoracks is one of the best-known nanosat startup companies offering CubeSat launches and tests platforms for earth observation and also includes biopharma research.

Data published in various reputed sources shows that the worth of the global space industry is estimated to be $350 billion and identified that it will surpass $550 billion by 2025. India contributes $7 billion which is around 2% of the global market. On a globally, about 17,000 small satellites are likely to be launched by 2030. Space missions were an exclusive playground for government space programs and for selected larger space enterprises. The whole scenario has been changed with the advancements and development in the success of student-developed small satellites opening the space industry drifting from educational Institutes/Universities to start ups. The market potential for small satellites is expanding with the rapid pace of technological advancements reducing the cost of satellite development. Universities and technology companies started space missions at an unprecedented pace. India is backed up with many small-scale players supporting the country's space missions and programs. The operations cost model for developing small satellites includes the cost of design, operations, and risk with regard to the dynamic complexity of the mission. A systems dynamic model can generate the cost model by working with Monte Carlo simulations which provide insight into the costs and risk aspects for operational designs [19].

According to the report published by the Times of India, there are 15 to 20 startups in the space technologies sector working on a variety of solutions to make payloads and launches more economical along with improving communications. One such Indian startup is a space technology company, Pixxel started in 2018 by two 20-year-old student entrepreneurs from BITS Pilani Institute with a vision to contribute meaningfully towards space exploration. Pixxel also reached out to potential customers in the areas of agricultural, mining, and petroleum sectors to identify how satellite imagery and data can be used to tackle inherent problems. Many of the entrepreneurs behind these firms are young space enthusiasts, who were a part of Isro's student satellite initiative YouthSat. Agnikul Cosmos incubated at IIT-Madras, IISc-incubated Bellatrix Aerospace, GrowX Ventures, and KA Enterprises are prominent startups working with ISRO in refining the products for space missions. Kawa Space is another prominent startup with the mission to build critical infrastructure that makes space-enabled solutions accessible to space enthusiasts. Bengaluru-based Astrogate, a Bangalore-based startup is working to replace traditional radio frequency systems with optical laser-based for satellite communications. Earth2Orbit is considered the first private space startup from India offering space launch advisory packed with consulting services. Exseed space is a startup that develops small satellite platforms and was the first startup that was selected by ISRO to build large satellites for Indian space missions. Dhruva Space, a Bangalore-based space company with the objective to expand spacecraft development private industry sectors in India. Dhruva in association with ISRO developed a follow-up satellite to HAMSAT-1 which was launched in 2005, catering to the societal needs in disaster management, amateur/emergency radio communication, and education. Dhruva Space provides application-agnostic space platform solutions where-in 300–1200 satellites will be sent to space every year spanning over the next 10 years. Bellatrix Aerospace is a Coimbatore-based research and development company that launch electric propulsion systems for satellites for dropping the satellite mission cost. Xovian is a New Delhi-based aerospace startup offering low-cost and sustainable solutions in satellite fabrication facilities. Skyroot aerospace, a Hyderabad-based SpaceTec startup founded by two former ISRO scientists. The mission is to build launch vehicles for small satellites. Recently, Azista BST Aerospace started in India based in Hyderabad to develop aerospace technologies and satellites. It carries out end-to-end design, development, fabrication, and integration of satellite systems including payloads and satellite-enabled ground systems, SATCOM, and hydrometeorology sensors and stations.

4 Summary

In this review, the current status of student satellite programs around the globe in general and India, in particular, is discussed. In addition, a survey of small satellite missions especially in the industry domain also is briefed. The advantages of more and more constellations of small satellites working together to accomplish a mission have greatly increased the spatial and temporary coverage of the earth. Having many satellites for a mission enables them to be used for various purposes, so that if some of the satellites fail to perform their intended task, we still have plenty in orbit to take over. Also, with small satellites, it is possible to make large data observations and build quick response capabilities with relatively low-cost technology. With the unique ability to cover any place on the earth, the constellations of small satellites could be more beneficial for many practical applications. Importantly, the recent Space Policy 2020 put forward by the Government of India emphasizes the role of private companies in the future space missions. To facilitate more activities and inventions in the aerospace sector, encouragement to student satellite projects is essential. Moreover, the possibility of potential development of quality human resources and start-up spinoffs are also the expected outcomes of the student satellite projects in India.

References

1. Burleigh SC, De Cola T, Morosi S, Jayousi S, Cianca E, Fuchs C (2019) From connectivity to advanced internet services: a comprehensive review of small satellites communications and networks. Wirel Comm Mobile Comput:17. Article ID 6243505. https://doi.org/10.1155/2019/6243505
2. Nielsen JFD, Larsen JA, Grunnet JD, Kragelund MN, Michelsen A, Sørensen KK (2009) AAUSAT-II, a Danish student satellite. ISAS Nyusu. https://vbn.aau.dk/files/19166464/www.aprsaf.org-feature-feature_71.html.pdf

3. Swartwout M (2011) AC 2011–1151: Significance of Student-built spacecraft design programmes it's impact on spacecraft engineering education over the last ten years
4. Poghosyan A, Golkar A (2017) CubeSat evolution: analyzing CubeSat capabilities for conducting science missions. Progr Aerosp Sci 88:59–83. https://www.sciencedirect.com/science/article/abs/pii/S0376042116300951
5. Prasad N (2015) India's small satellite mission: time for the next leap forward. Observer Research Foundation. https://www.researchgate.net/profile/Narayan_Prasad12/publication/305210211_India's_Small_Satellite_Mission_-_Time_for_the_Next_Leap_Forward/links/5784e2ec08aeca7daac57ceb/Indias-Small-Satellite-Mission-Time-for-the-Next-Leap-Forward.pdf
6. Angadi C, Manjiyani Z, Dixit C, Vigneswaran K, Avinash GS, Narendra PR, Prasad S, Ramavaram H, Mamatha RM, Karthik G, Arpan HV (2011) STUDSAT: India's first student Pico-satellite project. In: 2011 Aerospace conference. IEEE, pp 1–15. https://ieeexplore.ieee.org/document/5747469
7. Lamichhane K, Kiran M, Kannan T, Sahay D, Ranjith HG, Sandya S (2015) Embedded RTOS implementation for twin nano-satellite STUDSAT-2. In: IEEE metrology for aerospace (MetroAeroSpace), Benevento, pp 541–546. https://doi.org/10.1109/MetroAeroSpace.2015.7180715. https://ieeexplore.ieee.org/abstract/document/7180715/
8. Nagabhushana S, Dasiga S, Loganathan M, Rajulu B, Divya M (2014) Orbital analysis and hardware configuration for inter-satellite link in STUDSAT-2. In: 2014 IEEE aerospace conference. IEEE, pp 1–6. https://ieeexplore.ieee.org/abstract/document/6836349/?casa_token=IQydSZcRo9gAAAAA:HL0g1c_on14lKMvUd9vPAcx9DaWcZC9-mlu-ElbJxRQRD0SR_6AtEcVbZwb-XMRMHThTEFsEBXw
9. Murugan P (2020) Satellite projects by Indian students. Int J Eng Res Technol (IJERT) 9(3). https://www.academia.edu/download/64117189/satellite-projects-by-indian-students-IJERTV9IS030516.pdf
10. Prashanth G, Kumar A, Mulay S (2013) Pratham satellite: Faraday rotation based TEC measurement. http://nopr.niscair.res.in/bitstream/123456789/19415/1/IJRSP%2042(3)%20197-203.pdf
11. Kothari H, Lohiya P, Zubair A, Singh S (2020) Satellite structure of Advitiy (Second student satellite of IIT Bombay). In: Advances in small satellite technologies. Springer, Singapore, pp 535–544. https://doi.org/10.1007/978-981-15-1724-2_49
12. Mulay SS, Joshi J, Chati YS, Unhelkar VV, Bandyopadhyay S, Tamaskar S, Bommanahal M, Talnikar C, Kumar A, Hablani HB (2012) Attitude determination and control of Pratham, Indian Institute of Technology Bombay's first student satellite. Advances in the Astronautical Sciences, vol 145, pp 1509–1528. http://unhelkar.scripts.mit.edu/homepage/wp-content/uploads/2013/04/DyCoSS_SSKF_Pratham_Mar2012.pdf
13. Jha J, Thakur D, Neema K, Jadhav T, Rachh M, Bandyopadhyay S (2011) Design of the groundstation, polarization measurement setup and the social goal of Pratham, Indian Institute of Technology Bombay's first student satellite. http://citeseerx.ist.psu.edu/viewdoc/download?doi=10.1.1.731.5671&rep=rep1&type=pdf
14. Bandyopadhyay S, Jha J, Goel A, Thakur D, Neema K, Rachh M, Das S (2010) Measurement of total electron count of the ionosphere and the social goal of Pratham, Indian Institute of Technology Bombay's first student satellite. In: Proceedings of the International Astronautical Congress. http://publish.illinois.edu/saptarshibandyopadhyay/files/2020/02/Measurement-of-Total-Electron-Count-of-the-Ionosphere-and-the-Social-Goal-of-Pratham-Indian-Institute-of-Technology-Bombay%E2%80%99s-first-Student-Satellite.pdf
15. Sambhus MN, Kulkarni MR, Jakhotia MP, Kulkarni MS, Bondre MM, Chhajed MP, Marne MA, Chougule MO, Alalasundaram MS, Waghulde MD (2015) Project management and systems engineering of Swayam student satellite initiative. In: 66th International Astronautical Congress. https://www.researchgate.net/profile/Aniket_Marne/publication/296247328_PROJECT_MANAGEMENT_AND_IMPLEMENTATION_OF_HARDWARE_AND_SOFTWARE_INTERFACES_BETWEEN_SUBSYSTEMS_OF_SWAYAM_STUDENT_SATELLITE_INITIATIVE/links/57502e8908ae5c7e547a8c4a.pdf

16. Swartwout M (2018) Reliving 24 years in the next 12 minutes: a statistical and personal history of university-class satellites. https://digitalcommons.usu.edu/cgi/viewcontent.cgi?article=4277&context=smallsat
17. Xue Y, Li Y, Guang J, Zhang X, Guo J (2008) Small satellite remote sensing and applications–history, current and future. Int J Rem Sens 29(15):4339–4372
18. Lappas V, Kostopoulos V (2020) A survey on small satellite technologies and space missions for geodetic applications. In: Satellites missions and technologies for geosciences, p 123
19. Nag S, LeMoigne J, de Weck O (2014) Cost and risk analysis of small satellite constellations for earth observation. In: 2014 IEEE Aerospace conference. IEEE, pp 1–16
20. Hegde KM, Abhilash CR, Anirudh K, Kashyap P (2019, March) Design and development of RVSAT-1, a student nano-satellite with biological payload. In: IEEE aerospace conference. IEEE, pp 1–14. https://ieeexplore.ieee.org/abstract/document/8742113/?casa_token=YEAGdY-VHd0AAAAA:2ulZOGTlrX9aSXJRarkIbMsgzP1FEUpYWuWZqfVlNQCQcyo8ehaG4aemRi3U9dFIpQnqOP4kF10

Remote Sensing and Small Satellites

Analysis and Testing of Space Electronics' Package to Meet Launch Vehicle Vibration Loads

Santosh Joteppa, Shashanka M. Dibbi, and Vinod S. Chippalkatti

Abstract Satellite consists of many electronic packages, and during the launch of a satellite, it will be subjected to high random vibration loads which are generated by the launch vehicle. These electronic packages also known as modules, systems and/or subsystems will experience these vibration loads when mounted in the satellite and must be designed and qualified to meet these launch loads. These modules house the Printed Circuit Board (PCB) on which critical components such as HMCs, capacitors, CCGAs, flat packs and other leaded components, which are prone to vibration failures, need to be taken care during the design phase. Above all these components are high reliable and high cost items, this puts the constraint of first time right, and no failure is allowed. Hence, structural design and analysis become most important stage before manufacturing and testing the packages. This paper discusses one of the electronic packages designed for Indian communication satellite to meet such vibration loads and methods used to analyze the package virtually (through analysis software) before testing which saves the time and cost by reducing the number of prototypes. Three analyses (Eigen value analysis, quasi static analysis and random vibration analysis) have been carried out to qualify the module virtually by finite element method. Modal analysis has been carried out to determine the natural frequency and modal effective mass of the module. Quasi static analysis has been carried out to determine the stresses. Random vibration analysis has been carried out to see the acceleration levels and overall g_{rms} along with the stress levels experienced by the components. Based on the analyses, the qualification model (QM) was built and validated through testing after which the proto-flight model (PFM) and flight models were built.

Keywords Structural · Space · Random vibration · Electronics' package · Satellite

---

S. Joteppa (✉) · S. M. Dibbi · V. S. Chippalkatti
Design and Engineering, Centum Electronics Ltd., Bangalore, India
e-mail: santoshj@centumelectronics.com

S. M. Dibbi
e-mail: shashankd@centumelectronics.com

V. S. Chippalkatti
e-mail: vinod@centumelectronics.com

© The Author(s), under exclusive license to Springer Nature Singapore Pte Ltd. 2023
R.S. Priyadarshini and T. Sundararajan (eds.), *Advances in Small Satellite Technologies*, Lecture Notes in Mechanical Engineering, https://doi.org/10.1007/978-981-19-7474-8_4

1 Introduction

The electronic package consists of many mechanical parts, and the devices placed on the PCB are also made of mechanical parts. The primary function of these mechanical parts is to protect the electronics from vibration loads and to act as heat sink or provide path for heat transfer. These mechanical parts should provide sufficient stiffness to the PCB to withstand loads and also act as a heat sink to remove the heat from components and keep them within their operating temperature. These parts also serve other functions such as EMI/EMC, humidity protection. The packages must be cost-effective and easy to manufacture.

The design approach for the package is stiffness-based approach since the loads are dynamic in nature and weight is the critical parameter. The structural design details of the space grade electronic package to withstand vibration loads are discussed in detail in the paper.

2 System Description

The electronic package is mounted on the satellite panel. This panel will have a natural frequency around 50–60 Hz. Hence, the package has to be designed in such a way that the natural frequency of the package assembly is at least one octave more than the panel in order to avoid resonance and subsequently failure of the components.

This package houses two similar PCBs side by side in a single housing with electronic components mounted on it. The PCB is fixed to the housing using five screws. Three mounting holes are provided to fix the package on the satellite panel. The housing and the cover are made up of aluminum alloy AL 6061 T6/T651. The PCB is made up of FR-4 material and has two Hybrid Microcircuit (HMC) [1], pot core and two leaded capacitors which can be considered as critical components. The fasteners are stainless steel SS A2 70 grade.

The housing and cover are designed with 0.7 mm thick wall in order to reduce the weight, and ribs are added at various locations to add stiffness [2] to the parts. The cover is mounted with many number of screws to reduce EMI/EMC problems. The PCB is designed with four copper layers and is 1.6 mm thick. The exploded view is shown in Fig. 1. The total mass of the assembly is 900 g (Table 1).

3 Analysis

This section explains the details of design and analysis iterations that have been carried out for the package to meet the required stiffness. The unit should have the first fundamental frequency more than 120 Hz. The numerical code-finite element analysis (FEA) is used to study the dynamic behavior of package assembly design

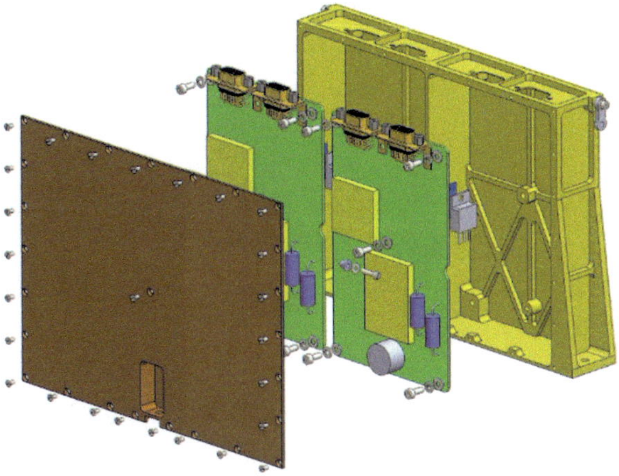

Fig. 1 Exploded CAD view

Table 1 Mass details

Components	Mass in grams
Housing	320
Cover	100
HMC A + HMC B (both two in numbers)	(24 + 16) each
Pot core (two in numbers)	12.3 each
Capacitors (four in numbers)	6.2 each
Connectors (four in numbers)	10 each
Fasteners	20
PCB with other components (two in numbers)	153.3
Total mass	900

using virtual prototype which is equivalent to flight or qualification model. The FEA results are analyzed thoroughly, the assembly is subjected to physical vibration test, and electrical parameters are measured pre- and post-vibration to verify the design. The unit is switched off during vibration.

FEM is a numerical procedure for solving the physical problems governed by a differential equation or an energy theorem, and it provides an approximate solution [3]. The finite element model of the assembly is shown in Fig. 2. All the critical components such as HMCs, capacitors and components having more than 5 g are modeled in the FEA (Table 2).

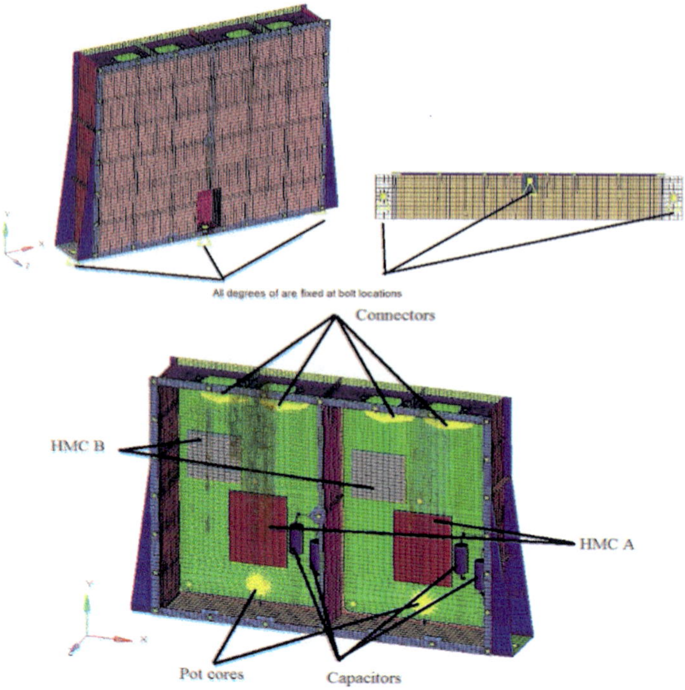

Fig. 2 FE model of assembly

Table 2 Number of elements and element types

Element type	Number of elements
Total number of nodes	35,530
Total number of elements	34,119
Total number of solid elements	4880
Total number of shell elements	28,986
Total number of beam elements	37
Total number of rigid elements	210
Total number of mass elements	6

3.1 Eigen Value Analysis

Eigen value analysis is carried out to determine the fundamental frequency of the package assembly. The first fundamental frequency of the assembly is found at **178 Hz** in finite element analysis. This meets the criteria of first fundamental frequency greater than 120 Hz (Fig. 3).

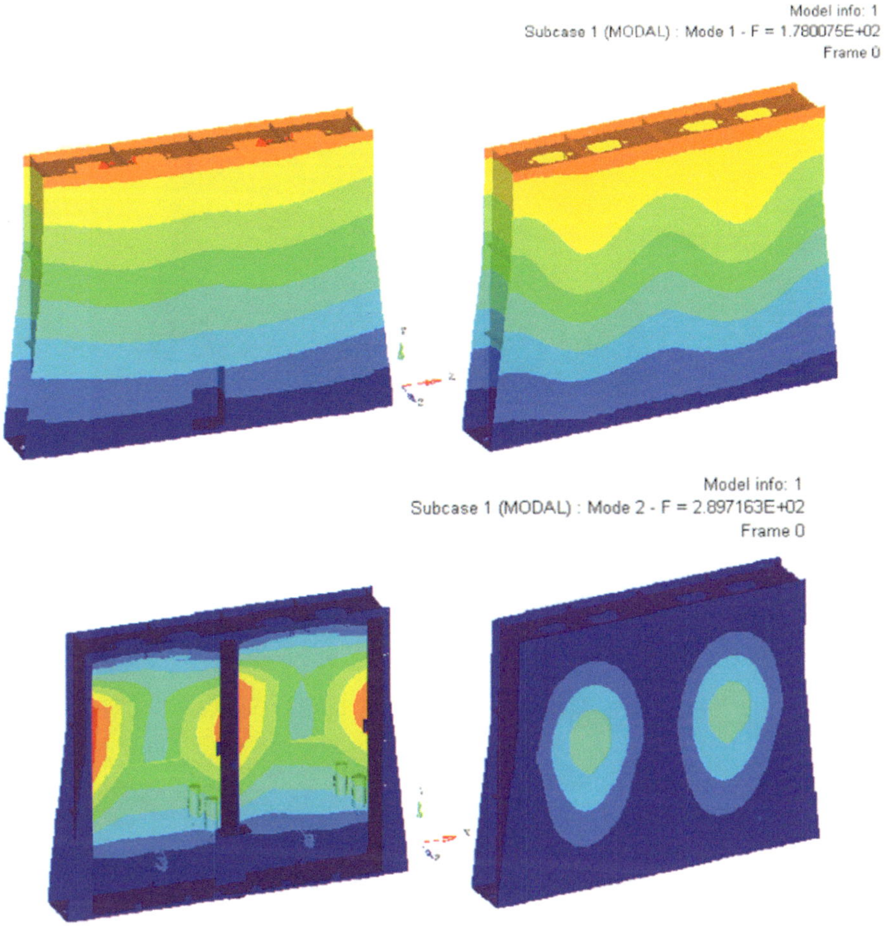

Fig. 3 Modal plots from the analysis

3.2 *Quasi Static Analysis*

Quasi static analysis is performed to simulate the constant acceleration in three axes, and the load levels are taken as:

(a) *X*-axis and *Z*-axis (in plane): 15 g
(b) *Y*-axis (out of plane): 30 g (Fig. 4).

It is observed that the stresses in the components are well within their allowable stresses, and the MOS is calculated considering a FOS of 1.1. MOS [4] is positive for all parts, which indicates that the parts are safe. The results are tabulated in Table 3.

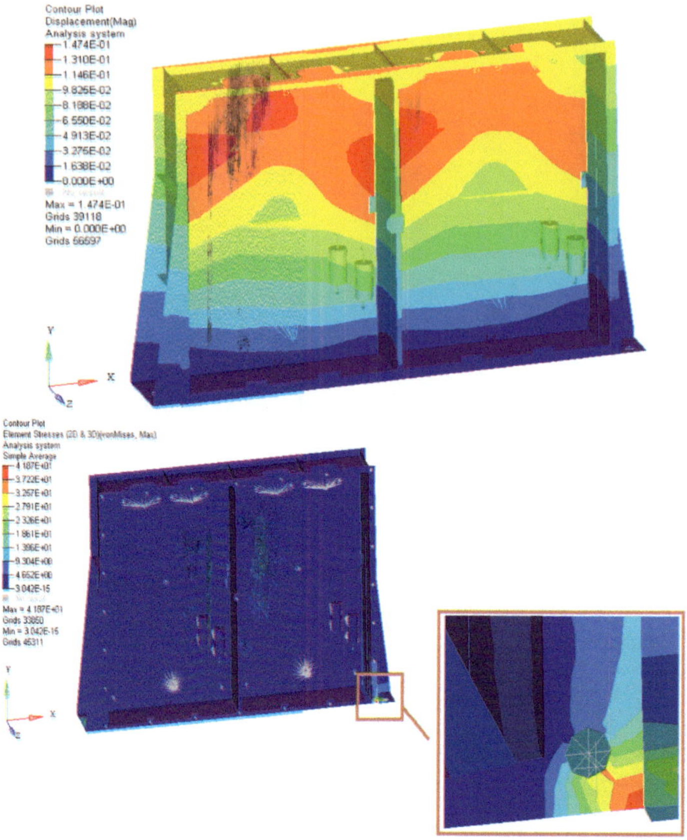

Fig. 4 Quasi static displacement and stress plots from the analysis

Table 3 MOS calculation

Sl. No.	Loading direction	von-Mises stress (MPa)	Yield stress (MPa)	FOS	MOS
1	*X*-direction	7.6 (at capacitor lead)	59	1.1	6.0
2	*Y*-direction	11.3 (at mounting bolt)	241	1.1	18.3
3	*Z*-direction	41.8 (at mounting bolt)	241	1.1	4.2

3.3 Random Vibration Analysis

Random vibration is the near real environment condition to which the package is subjected. This will give the insight into various aspects like design, workmanship and overall quality of the assembly.

Responses are taken on components and PCB, and the plots of frequency (Hz) versus PSD (g^2/Hz) are obtained (Fig. 5).

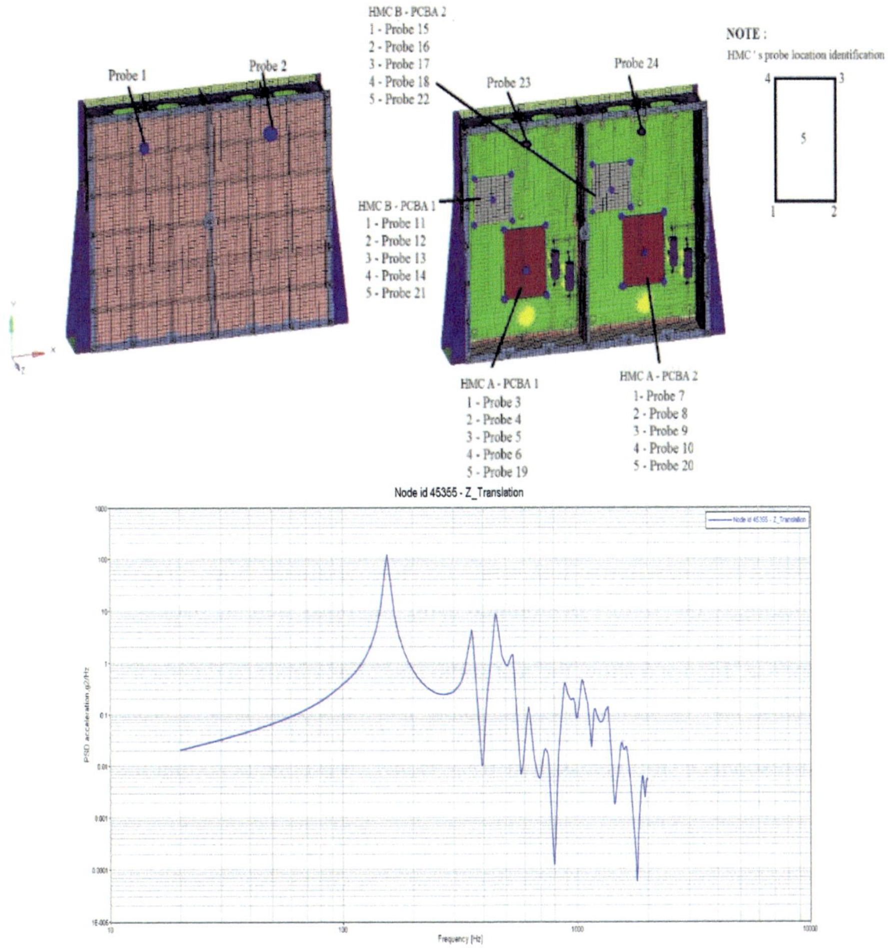

Fig. 5 Probes used to extract random responses and acceleration plot

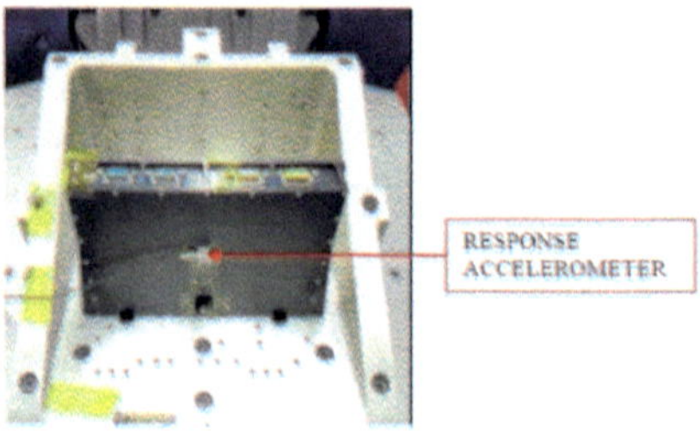

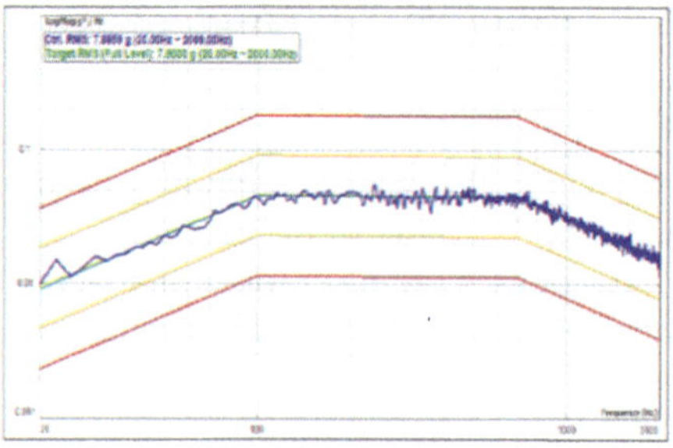

Random vibration profile

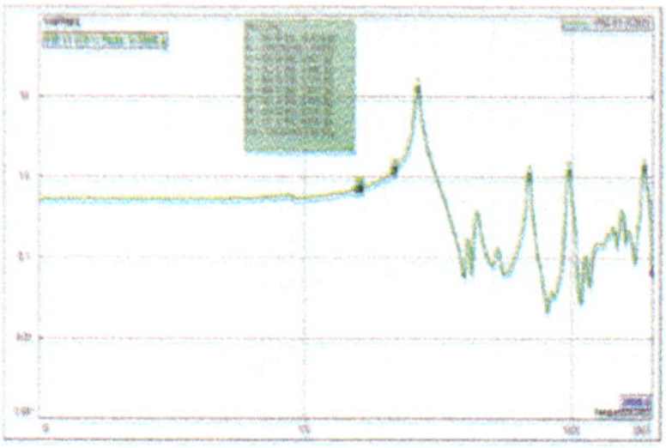

Pre Random Sine sweep

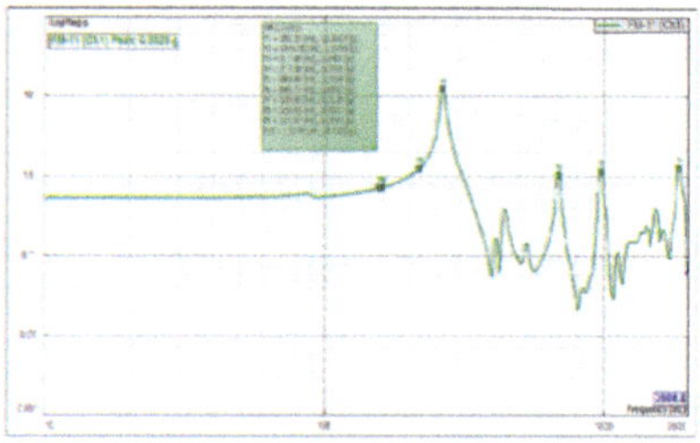

Post Random Sine sweep

Fig. 6 Test setup and acceleration responses

4 Results and Discussions

The qualified vibration fixture which does not have resonance up to 2000 Hz is used to test the package. The fixture is mounted using M8 bolts, and the package is mounted on the fixture with three M4 screws. An accelerometer is mounted on the package to take response. Figure 6 show the test setup and plots obtained during testing.

From testing, it is observed that the first fundamental frequency is at 262 Hz which is well above the required 120 and 178 Hz which is from analysis. The pre- and post-resonance shift is less than 10%. Hence, it is concluded that the package is structurally qualified.

5 Conclusion

From the above results of analysis and testing, it is observed that the fundamental frequency is above 120 Hz for the package assembly, and also, the stresses are under allowable material limits. Hence, the package is structurally qualified and is ready for use as flight model.

References

1. Carpenter data sheet for Kovar material
2. Steinberg DS. Vibration analysis for electronic equipment. A Wiley-Inter science Publications
3. Gokhale NS, Deshpande SS, Bedekar SV, Thite AN. Practical finite element analysis, by finite to infinite
4. Lyndon B. Johnson space center "Structural design requirements and factors of safety for spaceflight hardware"

Ku-Band Transmit–Receive Module for Imaging Radars

G. Ranjit Kumar Dora, Rajashekhar C. Biradar, and S. S. Rana

Abstract The Transmit–Receive module is one of the prime sub-systems required for designing an airborne or satellite imaging radar. This paper presents an overall detail of a highly optimized Ku-band Transmit–Receive module consisting of exciter and receiver sub-systems. The exciter receives the IF signal with necessary modulation like Linear Frequency Modulation (LFM) from signal processor module. The IF signal is upconverted to the transmit frequency using coherent LO signal and suitably amplified to the level required by the HPA. Suitable Ku-band calibration signal is also generated in this module for carrying out end-to-end calibration of the entire Transmit–Receive system. A common LO signal for both up-converter and down-converter is generated using a frequency synthesizer which is locked to high-stability Oven Controlled X-tal Oscillator (OCXO) for achieving high stability as well as extreme low phase noise. Various clock signals required for ADC/DAC/FPGA, etc. are also generated in this unit and are locked to the OCXO frequency. The received signal from antenna is amplified using a Low Noise Amplifier (LNA) after necessary protection with a limiter and downconverted to IF frequency using same LO used for up-conversion. The baseband section of the receiver provides necessary amplification to ensure optimum operation over the entire dynamic range. The signal is fed to the baseband data handling processor unit where it is digitized and processed. A compact Electronic Power Conditioner (EPC) is also housed along with the Transmit–Receive module for providing conditioned power supply. The paper briefly describes various options considered before freezing the final configuration. All the circuits are realized using Coplanar Waveguide (CPW) transmission lines on RT 6002, 10Mil substrate. To ensure stable and spurious free performance, the entire Transmit–Receive module is realized using several compartments to reduce

G. Ranjit Kumar Dora (✉) · R. C. Biradar
Centum Electronics Limited, Bangalore 560106, India
e-mail: ranjitkumard@centumelectronics.com

R. C. Biradar
e-mail: dir.ece@reva.edu.in

G. Ranjit Kumar Dora · S. S. Rana
School of ECE, REVA University, Bangalore 560064, India
e-mail: ssrana@centumelectronics.com

© The Author(s), under exclusive license to Springer Nature Singapore Pte Ltd. 2023

R.S. Priyadarshini and T. Sundararajan (eds.), *Advances in Small Satellite Technologies*,
Lecture Notes in Mechanical Engineering, https://doi.org/10.1007/978-981-19-7474-8_5

coupling among various stages. Special attention has been given to reduce the DC power requirement. Both thermal as well as structural issues are addressed while finalizing the mechanical housing. As the system is designed for satellite application, the paper also addresses various guidelines followed for component selection, fabrication processes, testing, qualification, etc. Various techniques adopted for the realization of low-cost/high-rel hardware are also addressed in the paper.

Keywords Transmitter · Receiver · Synthesizer · Ku-band · Imaging radars · Linear Frequency Modulation (LFM)

1 Introduction

The Transmit–Receive module is the major sub-system of the imaging radar and performs the following important functions:

- Up-conversion of the L-band Chirp signal centered at 1.75 GHz ± 125 MHz to Ku-band frequency with suitable power level.
- Generate bite signal at Ku-band for end-to-end evaluation of the radar system.
- Generate various LO/clock signals required by up/down-converters, ADCs, DAC, FPGA, etc. All these signals are phase locked to high-stability OCXO at 125 MHz.
- Amplify the signal received from antenna in a low noise front-end with necessary protection against strong signals entering the antenna as well as the transmitter power leakage. The receiver provides about 70 dB gain to boost the signal level to the level optimum for the ADC operation.

2 Design Methodology

2.1 *Transmitter and Synthesizer Module*

The Transmitter chain upconverts the L-band input signal to transmit frequency at Ku-band utilizing the LO signal (15.375 GHz) generated in synthesizer module. This Ku-band signal is further amplified and filtered and then fed to power amplifier. The block diagram of the up-converter module is shown in Fig. 1.

The synthesizer module consists of highly stable 125 MHz OCXO, which is used to generate various clocks and LO signals. Figure 2 shows the block schematics of the frequency generator having following outputs locked to 125 MHz OCXO. The main function of synthesizer module is generating the 1000 MHz clock signal for the processor, 15.375 GHz LO signal for the up-converter (Transmitter), 15.375 GHz LO signal for the down-converter (Receiver) and BITE signal generation at 17.125 GHz for checking the status of the system.

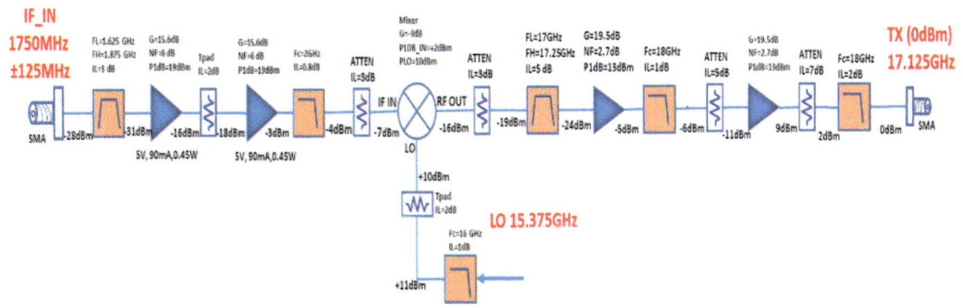

Fig. 1 Block diagram and chain budgeting of transmitter module

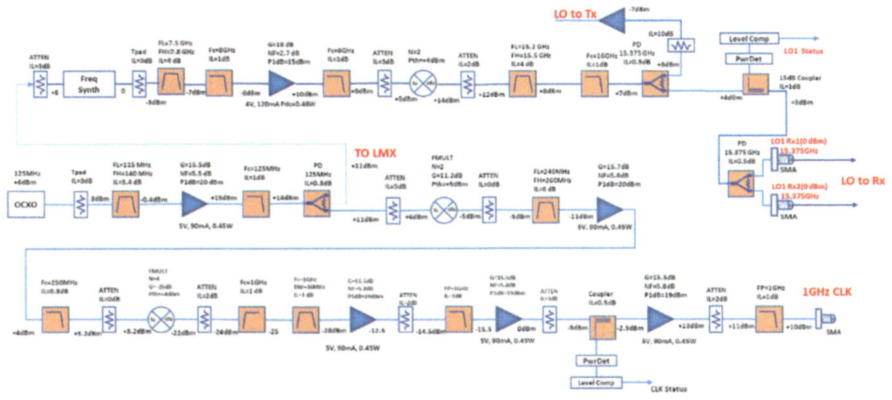

Fig. 2 Block diagram and chain budgeting of synthesizer module

As shown in the figure, the 125 MHz is used as reference to synthesize 7.6775 GHz frequency signal using a PLL-based synthesizer. This is followed by a frequency doubler to get 15.375 GHz LO signal for up/down-converters.

The 1000 MHz clock is derived from 125 MHz source by using X2 and X4 multipliers. Various amplifiers and filters required for the optimum performance are shown in the figure.

Table 1 gives detailed specification for the Transmitter and Synthesizer module.

Description of Transmitter Module

The chirp signal is generated in the processor module using high-speed DAC, 1750 MHz centered chirp signal with harmonics. BPF is used to pass the required band 1750 MHz ± 125 MHz and LPF is used in series to filter out all the other harmonics.

The Ku-band mixer is used to upconvert the IF signal, the 1750 MHz IF signal is fed to the IF port of the mixer and the LO port is fed with 15.375 GHz tone generated in the exciter module. The mixer IF and LO power levels are maintained to deliver optimum performance. Mixing of IF signal and LO signal will generate required

Table 1 Specifications of transmitter and synthesizer

Sl. No.	Parameters	Specifications
Transmitter		
1	IF input	L-band (1.750 GHz ± 125 MHz)
2	Input and output impedance	50 Ω
3	Input signal range	− 28 dBm
4	Tx nominal gain	30 dB ± 1.5 dB
5	Phase noise	− 75 dBc @100 Hz
6	Harmonic level and spurious	< − 50dBc (system requirement)
7	Tx output power level	0 dBm ± 1.5 dB
8	Tx frequency	Ku-band (17.125 GHz ± 125 MHz)
Synthesizer		
1	OCXO	125 MHz
2	OCXO phase noise	− 130 dBc/Hz @ 100 Hz offset
3	Clock frequency	1000 MHz
4	Clock power level	+ 10dBm ± 1 dB
5	LO frequency	15.375 GHz
6	LO power level	0 dBm ± 1.5 dBm
7	LO phase noise	− 80 dBc/Hz @100 Hz offset

Ku-band (1.75 GHz + 15.375 GHz = 17.125 GHz) upconverted signal. Microstrip bandpass filter is designed to pass the required Ku-band signal and followed by another Microstrip low-pass filter to remove harmonics and improve rejection.

Description of Synthesizer Module

The synthesizer module has the following functions:

- Generation of stable reference signal at 125 MHz with low phase noise.
- Clock for the processor module (1 GHz).
- 15.375 GHz LO generation for up/down-converters.

Figure 2 gives the block schematics of the exciter module showing clock generation, LO generation, up-conversion and bite generation.

Brief details of these functions are given below:

- **Generation of stable and low phase noise signal at 125 MHz**

To meet the stability and phase noise requirements, OCXO is chosen with phase noise of − 130 dBc/Hz at 100 Hz offset from carrier. OCXO output is filtered using 125 MHz BPF, and to meet the drive level of next stage, the signal is further amplified. The final filtered 125 MHz reference signal is distributed to two different circuits using two-way power dividers as shown in the figure. One output is used for generating 1000 MHz clock signal for processor (FPGA clock), and the other outputs are used for generation of LO signal at 15.375 GHz.

- **1 GHz clock for the processor module**

DAC requires low-jitter 1 GHz clock which is generated by multiplying the 125 MHz signal using frequency doubler and followed by frequency quadrupler. The 250 MHz signal from frequency doubler is fed to highly selective BPF where 250 MHz signal is passed, and other harmonics are rejected. The 250 MHz LC-BPF filter has a stopband attenuation of > 50 dB up to 1.4 GHz. Interdigital BPF topology is chosen to implement 1 GHz BPF; for wide stopband rejection in the next stage, LPF is used, and it provides rejection greater than 30 dB.

- **15.375 GHz LO generation**

For 15.375 GHz LO, the following two options are considered, using active/passive multipliers and using PLL-based frequency synthesizer.

The multiplier and frequency synthesizer approach both meet the phase noise performance required for the system performance; the synthesizer approach is preferred due to its compactness. However, the highest frequency available for space grade PLL-based frequency synthesizer with integrated VCO is 15 GHz. It was therefore decided to generate 7.6875 GHz frequency using frequency synthesizer and use a frequency doubler at the output to get 15.375 GHz. The output power level of synthesizer can be adjusted, and max power output from RF-out pin is 6dBm required for the next stage.

Final LO generation in this option has two approaches: use second harmonic of 7.6875 GHz from synthesizer and use frequency doubler after synthesizer (7.6875 GHz).

Active frequency multiplier is used after the synthesizer to get the required 15.375 GHz frequency. Necessary Microstrip filters are designed to have better rejection of 35 dB at Fo and better than 25 dB at 3 Fo.

Frequency multiplier has rejection of 35 dB at Fo and 25 dB at 3 Fo. The combination of both will therefore give about 60 dB harmonic rejection. There is provision to use 5 section Microstrip filter to improve the spurious/harmonic rejection in case the specifications are not met. The power level to amplifiers and frequency multipliers is adjusted using attenuator pads to meet the input power drive levels and to improve the return loss. The Transmitter and Synthesizer module is realized as shown in Fig. 3.

2.2 Receiver Module

The Receiver receives RF signal (17.125 GHz $\pm$ 125 MHz) from antenna and down-converts it to an IF signal of 1.75 GHz (L-Band). The limiter is used for LNA protection against any leakage power during transmission or strong spurious signals entering the antenna. This, being the first component of the receiver, plays a major

Fig. 3 Transmitter and synthesizer module

role in deciding the noise figure of the system. The bandpass filter is placed after LNA to filter out the harmonics and spurious signals. Mixer is a nonlinear device that is used to downconvert the input RF signal (Ku-band) to IF signal (L-band) by using Ku-band local oscillator. Adequate filtering is done at required frequency to remove spurious signals as well as lower/higher sidebands that are generated in the mixer. Digital attenuator is used in the IF section to adjust the overall receiver gain based on the requirements. The total gain in IF path is realized using several amplifying stages housed in separate enclosures to achieve stable behavior of the receiver. The block schematics of the receiver is given in Fig. 4.

Table 2 gives detailed specification for the Receiver module.

Description of Receiver Module

The receiver module is realized as shown in Fig. 5. The processor provides necessary controls signals required for the receiver modules. The power supply for the receiver modules is generated by using an external power supply unit. The detailed block diagram of receiver is shown in Fig. 4.

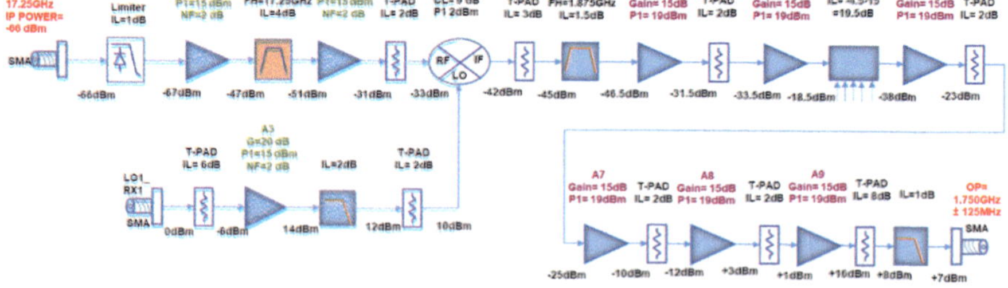

Fig. 4 Block diagram and chain budgeting of receiver module

Table 2 Specifications of receiver

Sl. No.	Parameter	Specifications
1	Input frequency	17.125 GHz ± 125 MHz
2	Receiver noise figure	6.5 dB (Max)
3	Nominal conversion gain	70 dB ± 1.5 dB with 15 dB attenuation set
4	Total gain flatness	± 1.5 dB over bandwidth at ambient
5	IF output frequency	1.75 GHz ± 125 MHz
6	Image rejection	< – 50 dB (Min)
7	Harmonics and spurious	< – 50 dB (Min)
8	Output P1dB point	+ 10 dBm (min)

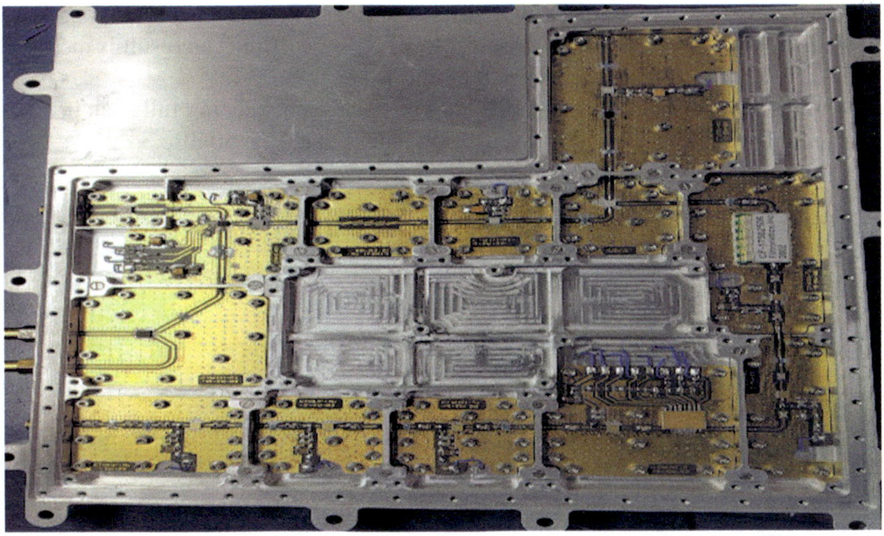

Fig. 5 Receiver module

Limiter is the first component of the receiver module. The limiter is basically used for LNA protection against any leakage power during transmission. LNA plays a major role to deciding the noise figure of the whole system. It is also proposed to use an SPDT switch following the limiter to remove any transients produced by the limiter during the leading edge of the high-power pulse. The switch is also used to inject the BITE signal at the receiver input for calibration/validation purpose. Two Ku band LNAs are used to achieve 40 dB gain which provides desired input level to drive the mixer. The bandpass filter (BPF) placed after LNA filters out the harmonics and spurious signals.

Mixer is a nonlinear device that is used to downconvert the input RF signal (17.125 GHz) to IF signal (1750 MHz) by using 15.375 GHz local oscillator which is generated by the exciter. Adequate filtering is done at IF frequency to remove spurious signals as well as lower/higher sidebands that are generated by the mixer.

A 5-bit digital attenuator (DA) is used at the IF level to adjust the overall receiver gain based on the requirements. The total gain in IF path is realized using six amplification stages for ensuring stable operation of the receiver over temperature variation as well as due to aging over the payload life.

3 Results

3.1 *Transmitter Measurements*

Figure 6 shows the gain flatness of 30 dB ± 1.5 dB and the transmitter outpour power of 0 dBm ± 1.5 dB with the input power level of − 28 dBm. The result is measured at 17.125 GHz ± 125 MHz with span of 250 MHz.

Figure 7 shows spurious and harmonics measurement keeping full span in signal analyzer. The harmonics and spurious level is less than < − 50 dBc.

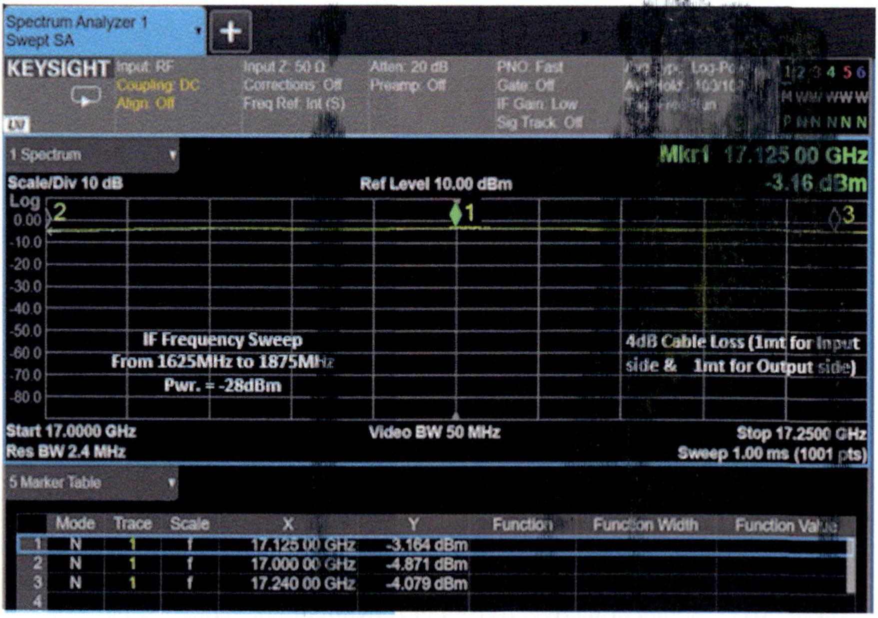

Fig. 6 Gain flatness of transmitter

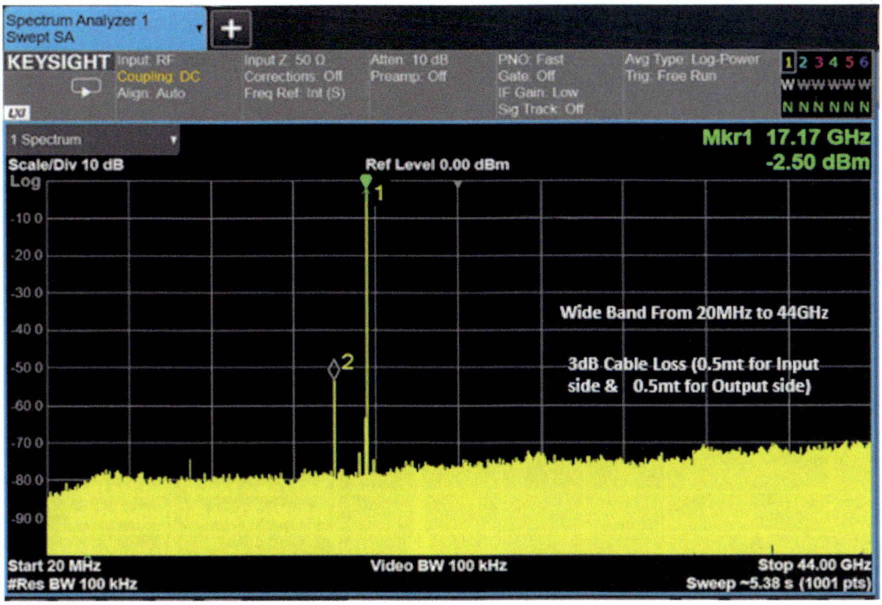

Fig. 7 Harmonics and spurious of transmitter

3.2 LO Signal Measurements

Figure 8 shows the output power level of LO signal, which is around 0 dBm ± 1 dB with respect to the input power level generated by OCXO.

Figure 9 shows the phase noise plot of LO signal -80dBc/Hz at 100 Hz offset.

3.3 Clock Signal Measurements

Figure 10 shows the output power level of clock signal, which is around + 10 dBm ± 1 dB with respect to the input power level generated by OCXO, and the harmonics and spurious level is less than < − 50 dBc.

3.4 Receiver Measurements

Figure 11 shows the gain flatness plots of receiver of 70 dB ± 1.5 dB with 15 dB attenuation set.

Figure 12 shows the noise figure measurement of receiver, which is less than 6.5 dB.

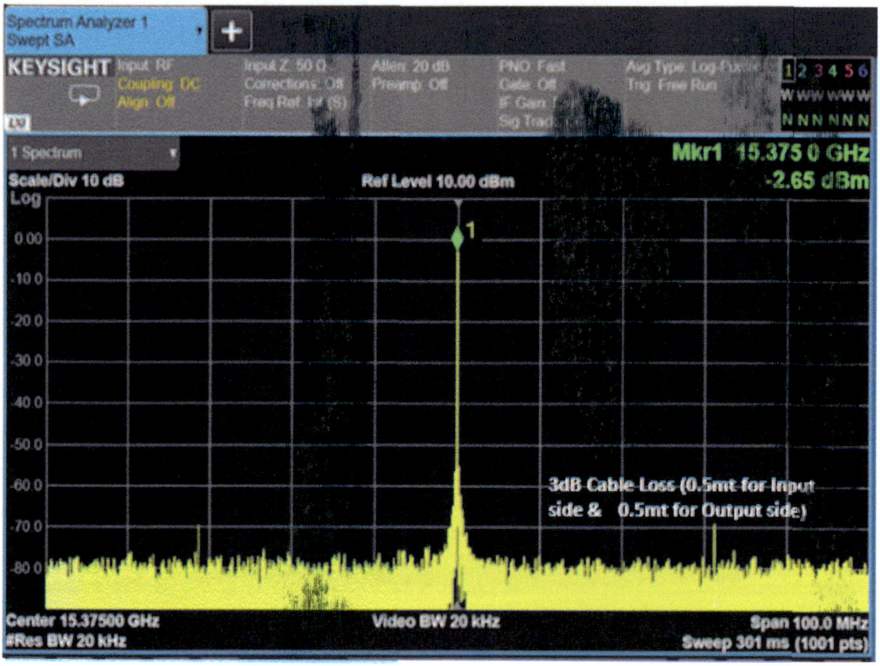

Fig. 8 Power measurement of LO signal

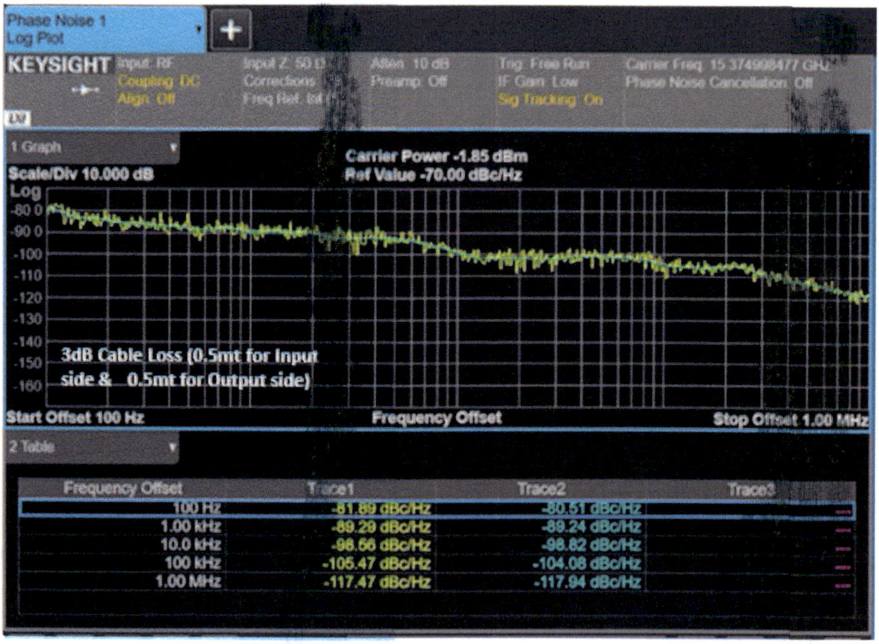

Fig. 9 Phase noise measurement of LO signal

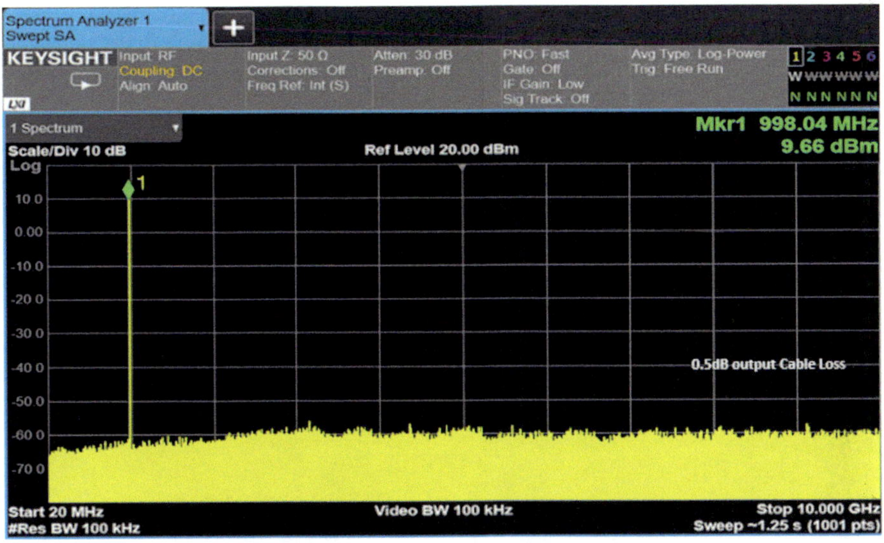

Fig. 10 Power measurement of clock signal

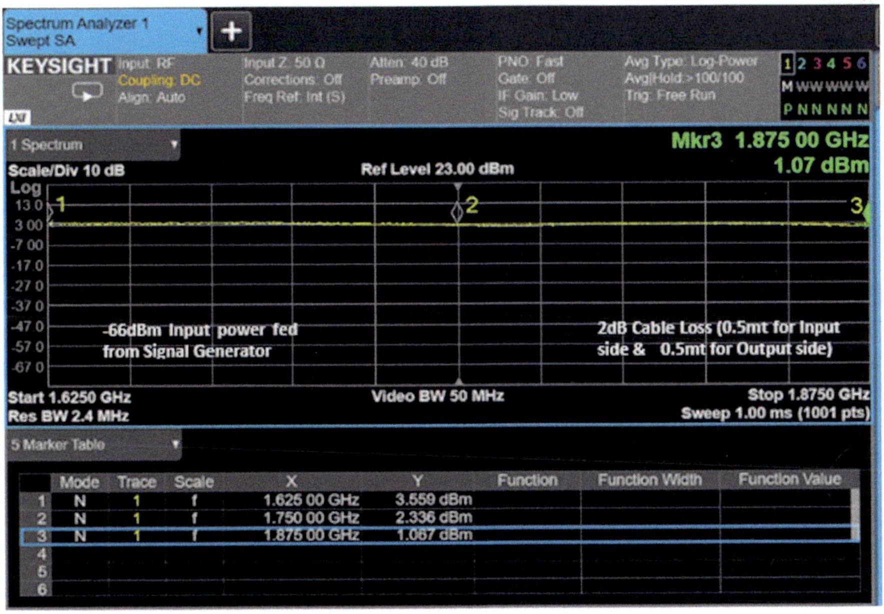

Fig. 11 Gain measurement plot of receiver

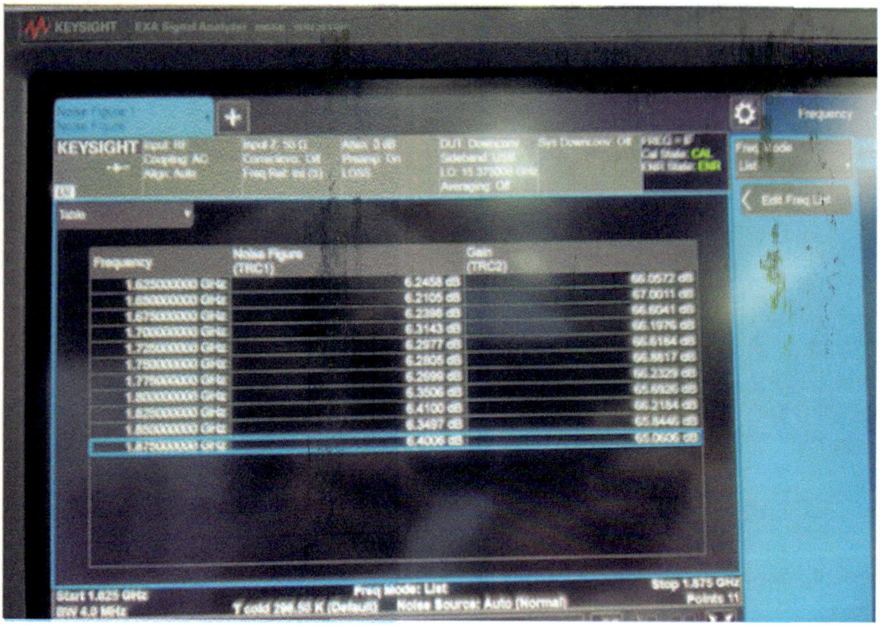

Fig. 12 Noise figure plot of receiver

4 Conclusion

The paper gives brief details of Transmit–Receive module consisting of Ku-band up-converter, coherent frequency generator giving 1000 MHz and 15.375 GHz and Ku-band low noise receiver. The coherent signals locked to 125 MHz OCXO show extremely low phase noise, and the Receiver has flat response over the full 250 MHz. The Ku-band Transmit–Receiver Module for imaging radars can be used for space application followed with the proper quality guidelines, such as Material Selection, Process Identification, Manufacture Procedure, Testing Procedure, Inspection Methodology and Qualification Procedure for fabricating QT model, and later stage can be used for flight model.

Small Object Detection in Remote Sensing Images

Melvin Kuriakose, P. S. Hrishikesh, Densen Puthussery, and C. V. Jiji

Abstract Automatic interpretation of remote sensing images is a fundamental but challenging problem in the field of aerial and satellite image analysis. It plays a vital role in a wide range of applications and is receiving significant attention in recent years. Even though many great progress has been made in this field, the detection of multi-scale objects, especially small objects in high-resolution satellite (HRS) and drone images, has not been adequately explored. As a result, detection performance both in terms of detection speed and accuracy turns out to be poor. To address this problem, we propose a convolutional neural network (CNN)-based single-stage object detector for the real-time and accurate recognition of remote sensing images. Our model predicts bounding boxes and corresponding class probabilities directly from images in a single assessment. This will result in a real-time object detection of images without compromising accuracy.

Keywords Small object detection · YOLO v4 · Remote sensing images

1 Introduction

In recent years, we have witnessed a significant advancement in the progress of remote sensing technologies. This results in a large amount of high-quality satellite and aerial images for research and investigation. Object detection in remote sensing images (RSIs) determines whether a given aerial or satellite image contains one or more

M. Kuriakose (✉) · P. S. Hrishikesh · D. Puthussery · C. V. Jiji
College of Engineering, Trivandrum, India
e-mail: memelvin@cet.ac.in

P. S. Hrishikesh
e-mail: hrishikeshps@cet.ac.in

D. Puthussery
e-mail: puthusserydenson@cet.ac.in

C. V. Jiji
e-mail: jijicv@cet.ac.in

© The Author(s), under exclusive license to Springer Nature Singapore Pte Ltd. 2023

R.S. Priyadarshini and T. Sundararajan (eds.), *Advances in Small Satellite Technologies*, Lecture Notes in Mechanical Engineering, https://doi.org/10.1007/978-981-19-7474-8_6

objects belonging to the class of interest and locates the position of each predicted object in the image. Being an elemental problem in the field of aerial and satellite image analysis, it has extensive applications, especially in disaster management, resource monitoring, road traffic analysis, military applications, etc.

The existing object detection models can be divided into two categories: one is a two-stage detector like faster R-CNN [1] and the other is the single-stage detector such as SSD [2] and YOLO [3]. In two-stage detectors, the first stage is to generate potential bounding boxes in an image and then run a classifier on these proposed boxes. Those complex pipelines are slow and difficult to optimize. On the other hand, for the single-stage detectors, a single neural network predicts bounding boxes and corresponding class probabilities directly from full images in a single assessment.

In this paper, we propose our model based on YOLO v4 [4] for the small object detection in RSIs. Since the whole detection pipeline of You Only Look Once (YOLO) is a single neural network, it can be optimized in a comprehensive manner. YOLO v4 achieved state-of-the-art results for detection of MS COCO [5] dataset with 43.5% average precision (AP) and running at a real-time speed of 65 frames per second (FPS) with Tesla V100 GPU.

2 Methodology

2.1 Proposed Method

The proposed model (Fig. 1) consists of three main components, a backbone for the feature extraction, neck for the feature fusion, and finally, the head for predicting the bounding box coordinates, object classes, and objectness. Our model uses CSPDarkNet 53 as the backbone. CSPDarkNet 53 is a combination of DarkNet 53 [6] and Cross-Stage Partial Network (CSPNet) [7] (Fig. 2).

DarkNet 53 consists of 53 convolutional layers mainly with 3×3 and 1×1 filters along with shortcut connections. The DarkNet 53 model is initially pretrained on ImageNet [8] dataset. This pretrained model extracts both low- and high-frequency details from the input image. The CSPNet helps in efficient feature extraction while reducing the amount of computation. The CSPNet follows an architecture as shown in Fig. 2. The output from the CSPDarkNet 53 is given as the input to the Dense Connection (DC) block [9] with four dense units consisting of 3×3 and 1×1 densely connected convolutional layers as shown in Fig. 1. The DC sequentially expands the feature maps to 256, 512, 512, and 512, respectively. This strengthens the feature extraction, and these feature maps are concatenated to form 2304 feature maps. The final output DC block is reduced to 1024 channels using a convolutional layer with a kernel of size 3×3.

In the proposed method, the Spatial Pyramid Pooling (SPP) block [9] forms the neck of the detector. It comprises three max-pooling layers for the concatenation of local region features that are extracted and converged by multi-scale pooling. In order

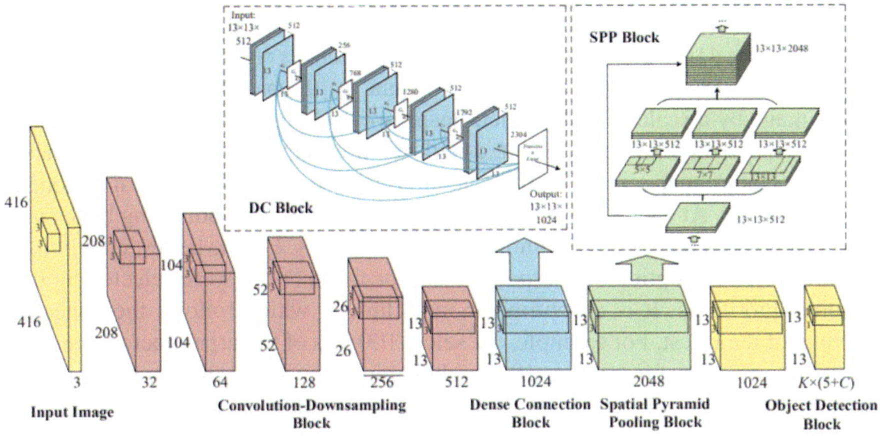

Fig. 1 YOLO model architecture

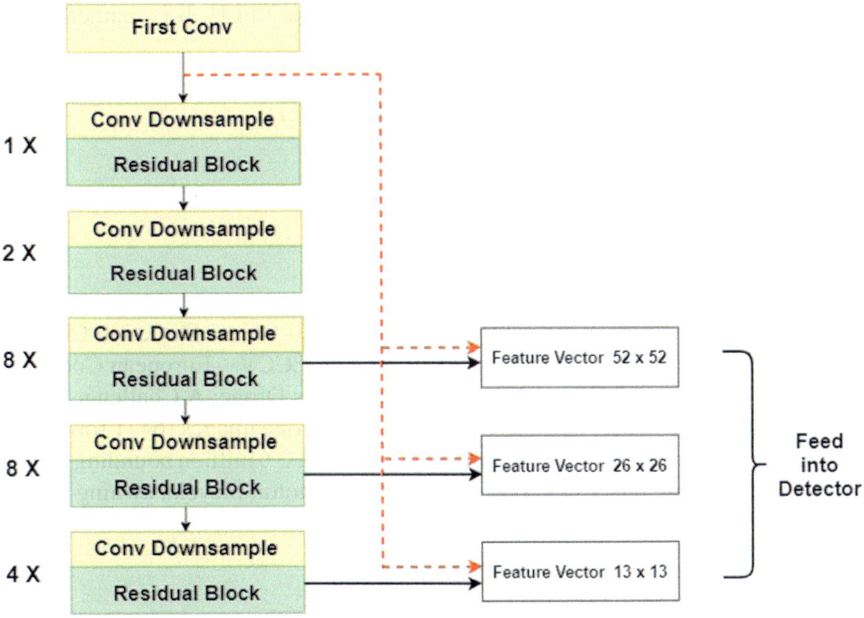

Fig. 2 CSPDarkNet 53 backbone

to reduce the number of input feature maps from 1024 to 512, the 1×1 convolution is adopted before the pooling layer. The bounding boxes and corresponding class probabilities are finally predicted in the object detection block, in which the high-resolution output feature maps from DC block are reconstructed and concatenated with the low-resolution feature maps of the SPP block.

Our model makes predictions on three sets of feature map sizes. That is when an input of 416×416 is given that the detection layer makes predictions on feature maps of sizes 13×13, 26×26, and 52×52. Primarily, the backbone downsamples the input image by a scale of 32. Further, layers are upsampled by up-sampling factors 2 & 4 respectively, and concatenated with feature maps of previous layers having identical feature map sizes. Finally, detections are made on all three scales using the detection layer. At each scale, each of the cells predicts three bounding boxes using three anchors, making the total number of nine anchors. For the better recognition of small objects, the optimal hyperparameters for the network were selected by using genetic algorithms. It is an optimization technique which follows the concept of survival of the fittest. For example, we select 100 sets of hyperparameters randomly. Then, we use them for training 100 models. Later, we select the top ten performed models. For each selected model, we create ten slightly mutated hyperparameters according to its original. We retrain the models with the new hyperparameters and select the best models again. As we keep the iterations, we should find the best set of hyperparameters. In addition to this, our model used the Mish [10] activation function for the CSPDarkNet 53 backbone and detector block. Previous studies show that Mish outperforms the conventional activation functions like ReLU. It is mathematically defined as:

$$f(x) = x * \tanh\left(\ln\left(1 + e^{x}\right)\right) \quad (1)$$

2.2 Dataset Description

For the proposed model, we used the VisDrone 2020 Object Detection Challenge [11] dataset. The challenge was in conjunction with ECCV—European Conference on Computer Vision—2020 workshop: "Vision Meets Drone: A Challenge".

The benchmark dataset consists of 10,209 static drone images (6471 for training, 548 for validation, and 3190 for testing) with more than 2.6 million bounding boxes of ten classes: car, van, bus, truck, pedestrian, person, motor, bicycle, awning-tricycle, and tricycle.

2.3 Experimental Setup

We used 24,000 steps to train the model. The step decay learning rate scheduling strategy is adopted with an initial learning rate of 0.01 and a multiplication factor of 0.1 at the 19,200 steps and the 21,600 steps, respectively. For training, we used a batch size of 64 and an image size of 1024×1024. For testing, batch size of 1 and the image size of 1664×1664 were used. Tesla P100 GPU using DarkNet library was used for the whole experiment.

3 Evaluation

3.1 Evaluation Metrics and Loss Function

As illustrated in Fig. 3, precision, recall, and IoU, Intersection over Union, can be used for evaluating an object detection model. IoU loss is used as the loss function, which takes into consideration the area of the predicted bounding box (BBox) and the ground truth BBox. This idea was improved furthermore by GIoU—Generalized Intersection over Union—loss, in which we find the smallest area BBox that can simultaneously cover the predicted BBox and ground truth BBox and use this BBox as the denominator to replace the denominator originally used in IoU loss. This will include the shape and orientation of an object in addition to the coverage area.

The mean of the average precision (mAP) is used for the evaluation. Specifically, AP, APIoU = 0.50, APIoU = 0.75, ARmax = 1, ARmax = 10, ARmax = 100, and ARmax = 500 metrics were used to evaluate the results of detection algorithms as described in Table 1.

All metrics are computed allowing for at most 500 top-scoring detections per image (across all categories). These criteria will penalize the missing and duplicate detections of the objects.

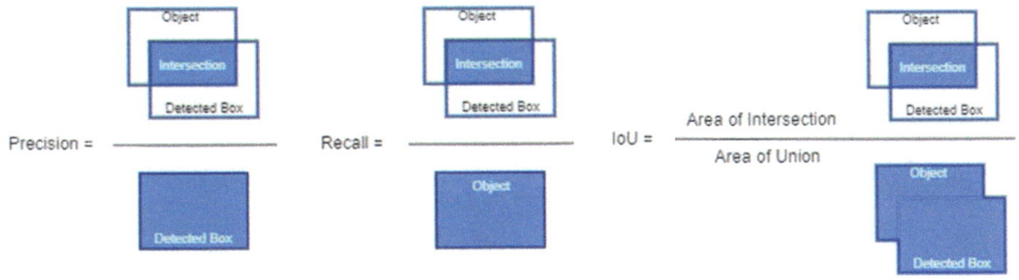

Fig. 3 Illustration of precision, recall, and intersection over union

Table 1 Evaluation metrics used for the VisDrone object detection challenge

Measure	Perfect (%)	Description
AP	100	The average precision overall ten IoU threshold values (i.e., [0.5:0.05:0.95]) of all object categories
APIoU = *i*	100	The average precision overall object categories when the IoU overlaps with ground truth are larger than *i*, for *i* = 0.50 and 0.75
ARmax = *j*	100	The maximum recall given *j* detection per image, for *j* = 1, 10, 100, and 500

4 Result

4.1 *Result Analysis*

The proposed method was efficient in recognizing very small objects like pedestrians and occluded objects in the images and achieved an mAP score of 22.7 in the VisDrone object detection evaluation metrics.

Figure 4 shows an example of the detection result from the model. The benchmark mAP score for the object detection challenge was 7.88, and we obtained comparable results with the top-performing teams as depicted in Table 2. DroneEye2020 which is a three-stage detector based on Cascade R-CNN was the top-performing model with a mAP score of 35.57. Our result can be further improved by using an ensemble of the top-performing feature extractor models like CSPDarkNet 53, ResNeSt 101 [12], and HRNet-W40 [13]. Also, by increasing the image size from 1024 $\times$ 1024 to 1664 $\times$ 1664 during the training will result in better model performance. But, this will compromise the computational costs.

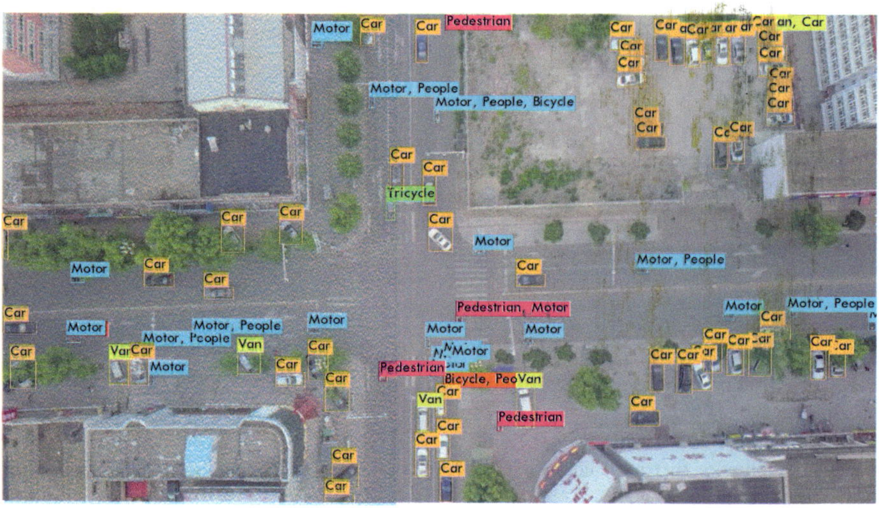

Fig. 4 Detection result from our model

Table 2 Performance comparison of our proposed method

Team	Method	mAP
Team 1	DroneEye2020	34.57
Team 2	TAUN	34.54
Team 3	CDNet	34.19
…	…	…
Our team	YOLO v4	22.70

5 Conclusion

In this paper, we proposed an object detection model based on YOLO v4 for the small object detection in satellite and drone images. Since the images are captured from very high altitudes, the network parameters of our model were optimized to recognize very small objects. Genetic algorithms were used for this purpose. Additionally, Mish activation function was used for better performance. The proposed method has achieved comparable results with top-performing teams in VisDrone 2020 object detection challenge and obtained an mAP score of 22.70.

References

1. Ren S, He K, Girshick R, Sun J (2017) Faster RCNN: towards real-time object detection with region proposal networks. In: IEEE transactions on pattern analysis and machine intelligence
2. Liu W, Anguelov D, Erhan D, Szegedy C, Reed S, Fu C-Y, Berg AC (2016) SSD: single shot multibox detector. In: European conference on computer vision
3. Redmon J, Divvala S, Girshick R, Farhadi A (2016) You Only Look Once: unified, real-time object detection. In: IEEE conference on computer vision and pattern recognition
4. Bochkovskiy A, Wang CY, Liao H (2020) Yolov4: optimal speed and accuracy of object detection. arXiv:2004.10934
5. Lin T-Y, Maire M, Belongie S, Hays J, Perona P, Ramanan D, Dollar P, Zitnick CL (2014) Microsoft COCO: common objects in context. In: European conference on computer vision
6. Redmon J, Farhadi A (2016) YOLOv3: an incremental improvement. arXiv:1804.02767
7. Wang C, Liao HYM, Wu Y, Chen P, Hsieh J, Yeh H (2020) CSPNet: a new backbone that can enhance learning capability of CNN. In: IEEE/CVF conference on computer vision and pattern recognition workshops
8. Russakovsky O, Deng J, Su H, Krause J, Satheesh S, Ma S, Huang Z, Karpathy A, Khosla A, Bernstein M, Berg AC, Fei-Fei L (2015) ImageNet large scale visual recognition challenge. Int J Comp Vis
9. Huang Z, Wang J (2019) DC-SPP-YOLO: dense connection and spatial pyramid pooling based YOLO for object detection. arXiv:1903.08589
10. Misra D (2020) Mish: a self regularized non-monotonic activation function. arXiv:1908.086
11. Zhu P, Wen L, Dawei D, Bian X, Qinghua H, Ling H (2020) Vision meets drones: past, present and future, arXiv:2001.06303
12. Zhang H, Wu C, Zhang Z, Zhu Y, Zhang Z, Lin H, Sun Y, He T, Mueller J, Manmatha R, Li M. A. Smola.: ResNeSt: split-attention networks. arXiv:2004.08955
13. Sun K, Zhao Y, Jiang B, Cheng T, Xiao B, Liu D, Mu Y, Wang X, Liu W, Wang J. High-resolution representations for labeling pixels and regions. arxiv.1908.07919

Orbital Platform and Payload Experiments

PS4-Orbital Platform: An Ideal Suite for Scientific Experiments

U. Sanood, M. Sandeep, Aaron Bapista, C. G. Suresh Nair, M. J. Lal, and S. R. Biju

Abstract PS4-Orbital Platform (PS4-OP) refers to a novel idea by ISRO (Indian Space Research Organisation) to use the spent PS4 stage (fourth stage of PSLV) as a three-axis-stabilized platform for small scientific payloads to carry out in-orbit scientific experiments for an extended duration of 4–6 months. The PS4 stage has standard interfaces and packages for power generation, telemetry, tele-command, stabilization, orbit keeping and orbit maneuvering. The scientific community/research organizations can design the scientific payload and utilize the OP interfaces offered by spent PS4 stage for powering, data management and specific experimental requirements. In short, ISRO is extending its expertise in space technology to the scientific community as a platform to design, develop and validate their experiments in an effective manner.

Keywords Orbital Platform · Attitude control · Thrusters

1 Introduction

With increasing demand for carrying out scientific experiments in outer space, ISRO aims to convert the fourth stage of Polar Satellite Launch Vehicle (PSLV) into an Orbital Platform. PS4-Orbital Platform (PS4-OP) is a novel concept by ISRO to use the spent PS4 stage for an extended duration of 4–6 months in a Low Earth Orbit, which otherwise would end up as space debris immediately after the mission objective.

The Orbital Platform has standard interfaces and packages for power generation, telemetry and tele-command. Power is generated using flexible solar panel mounted around PS4 propellant tank in line with any large-scale satellite. An independent Navigation, Guidance and Control (NGC) chain with dedicated avionics and cold gas Reaction Control System (RCS) thrusters are used to stabilize the platform in 3 axes. The left out Helium gas pressurant in the pressure vessel after the main mission

U. Sanood (✉) · M. Sandeep · A. Bapista · C. G. Suresh Nair · M. J. Lal · S. R. Biju
Vikram Sarabhai Space Centre, Thiruvananthapuram 695022, India
e-mail: meetsanu85@gmail.com

© The Author(s), under exclusive license to Springer Nature Singapore Pte Ltd. 2023

R.S. Priyadarshini and T. Sundararajan (eds.), *Advances in Small Satellite Technologies*, Lecture Notes in Mechanical Engineering, https://doi.org/10.1007/978-981-19-7474-8_7

is used as the working medium for the cold gas thrusters. PS4-Orbital Platform NGC chain would be activated after the completion of the main mission objectives.

Orbital Platform supports different functional modes, namely "Sun pointing," "Earth pointing" and "Desired attitude," which can be configured through telecommand (commands issued from ground stations). By default, PS4-OP will be in "sun pointing" mode, with solar panels oriented toward the sun for effective power generation. As per the PS4-OP mission requirements, any axis of PS4-OP can be aligned toward earth (Earth pointing mode) or PS4-OP can be oriented to any desired attitude as per the requirement of the scientific payload if any.

The advantage of this concept is that scientific community/research organizations can design the experimental payload utilizing the standard interfaces of Orbital Platform, without being concerned about requirements like power, Navigation, attitude control, data management and specific requirements including tele-command. The turnaround time in this scenario for the scientific payloads would be less, reduce costs and gives more emphasis on the scientific experiment rather than auxiliary needs for the payload sustenance in space. In short, ISRO is extending its expertise in space technology to the scientific community as a platform to design, develop and validate their experiments in an effective manner, which otherwise could have been possible only through dedicated nano/micro satellite missions [1, 2]. PS4-Orbital Platform thus eliminates the extra efforts, cost and time required for customers to convert their payload into individual satellites.

In a phased manner, the various features of the OP conceived were demonstrated. As part of demonstration of the concept, the spent PS4 stage was maintained for 10 orbits as a stabilized platform in PSLV-C38 mission. In 2nd phase of the Orbital Platform experiment, power generation through solar panel, telemetry and data downlink for the payloads housed in OP had been demonstrated in PSLV-C45 mission. Data from three scientific payloads were available for duration of 2 months. As 3rd phase of the Orbital Platform experiment, it is planned to demonstrate the three-axis attitude stabilization and tele-commanding capability, in one of the immediate PSLV missions.

The rest of the paper is organized as follows. Section 2 discusses the configuration of the Orbital Platform. Section 3 discusses the details of 2nd phase of PS4-OP experiment. Opportunities for future work and concluding remarks are presented in Sect. 4.

2 Sub-system Configuration

Following are the major elements of the Orbital Platform:

- NGC System
- Attitude control system
- Power system
- Telemetry and Tele-command system.

2.1 NGC System

The PS4-OP Avionics configuration is planned as a separate independent system without affecting the functional NGC chain. This is done to ensure that failure of PS4-OP elements is not propagated to functional NGC System. PS4-OP Avionics is configured as a single chain configuration, with all associated elements interconnected through an exclusive MIL 1553 bus (Fig. 1).

Indigenous Vikram 1601 based Mission Computer is used as the bus controller, and does the function of Navigation, Guidance and Control which are executed in required periodicity for the attitude stabilization and pointing requirements demanded by the payload.

Exclusive NavIC (Navigation using Indian Constellation) Receiver provides the NavIC/GPS (Global Positioning System) based position information to Mission Computer. Rates are provided by MEMS Rate Gyro package. In addition, absolute attitude information is derived using 4 No's of Micro Coarse Analog Sun sensor (MCASS) and a Magnetometer. The analog outputs of MCASS and Magnetometer are acquired through Data acquisition unit with Sigma-delta Analog to Digital

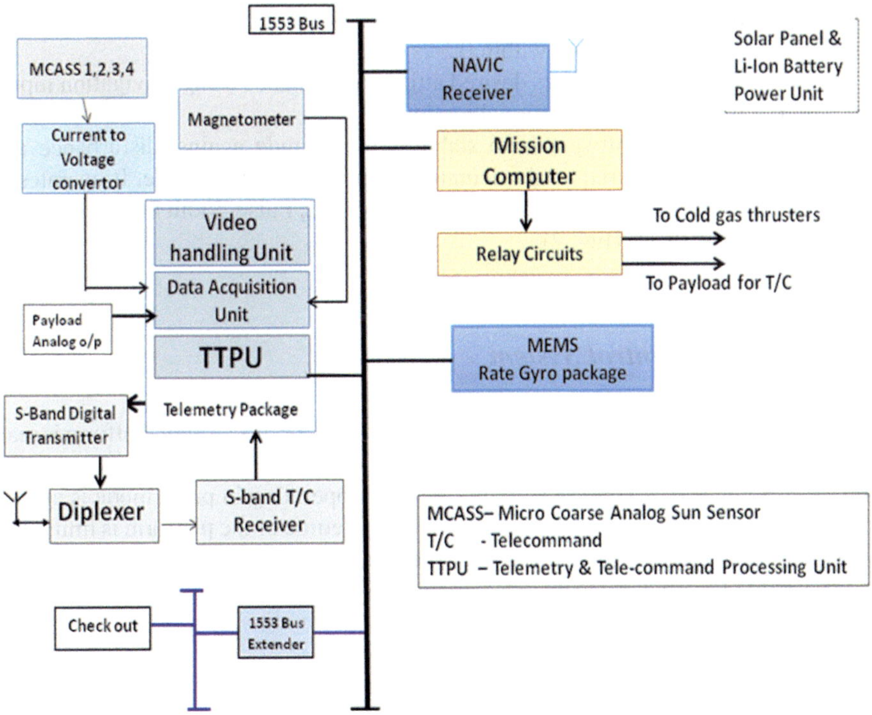

Fig. 1 PS4-OP NGC architecture

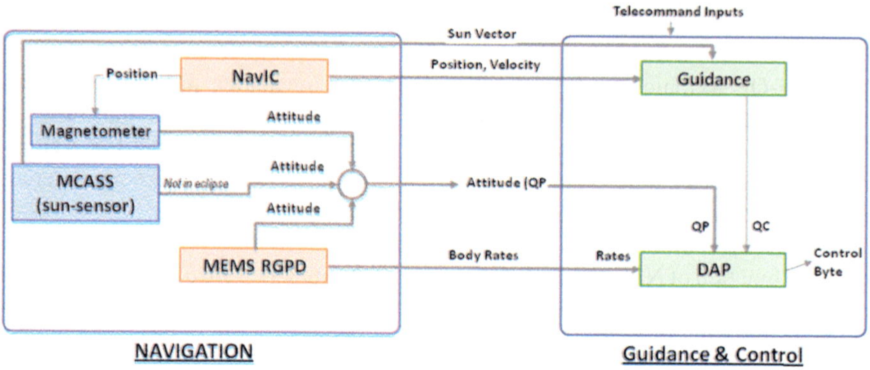

Fig. 2 Software overview

Convertors. 4 Nos of MCASS sensors are mounted to get the full coverage of the sun in any orientation. Each MCASS sensor has a range of ± 60°.

Navigation software computes the current orientation from Sun sensor, magnetometer and rate data, using a 6-state Kalman filter. This configuration can ensure pointing accuracy within ± 1° during sunlit period and ± 5° during eclipse (or best possible) and variation in rate within 0.5°/s during stabilized regime.

Guidance software generates desired attitude commands using Navigation inputs as required by respective operational mode.

Digital Autopilot software is to stabilize the attitude against disturbance and track the vehicle in the required orientation as per operational mode. It operates in different modes: 1. Pulse Width Modulation (PWM) 2. Pulse Width Pulse Frequency Modulation (PWPFM) (Fig. 2).

2.2 *Attitude Control System*

The 3-axis attitude stabilization (Pitch, Yaw and Roll) for this Orbital Platform is realized through a separate control power plant which is independent from the primary mission. 8 nos of 1 N Helium Cold gas thrusters operating in pulse mode is chosen as control power plant for Orbital Platform. The lifetime of the platform is limited by the quantity of Helium gas left out after the primary PS4 stage operation. Thrusters are assembled around the Vehicle Equipment bay, using additional brackets (Fig. 3).

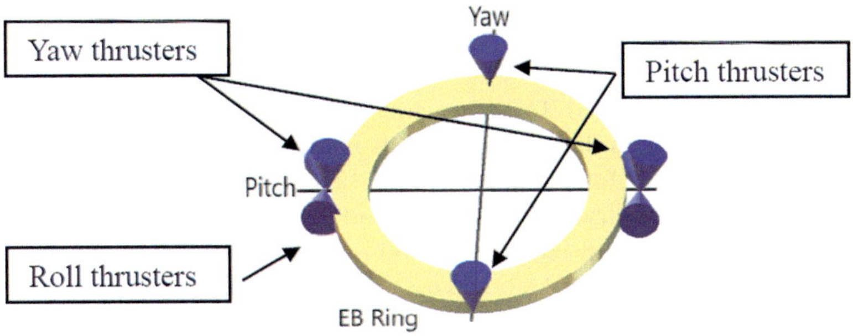

Fig. 3 Cold gas thruster configuration

2.3 Power System

The PS4-OP Avionics packages are powered by Flexible Solar Panel (FSP) in conjunction with 50AH Li-ion battery in battery tied configuration. Power conditioning unit housing the necessary circuits to interface the Solar panel, battery, packages and payloads (Fig. 4).

The solar panel for PS4 stage is designed as a flexible Carbon Fiber Reinforced Plastic (CFRP) laminate. The laminates are assembled as two separate halves around the PS4 tank. The laminates are mounted on CFRP sandwich ribs that bridge the fore end and aft end flanges of the PS4 propellant tank. The FSP laminate is fabricated using composite structures. Each laminate is mounted on 5 nos. of CFRP sandwich ribs.

The proposed power generation scheme was implemented in PSLV-C45. As 1/3rd area of solar panel is only exposed to sun at any instant, the power generated is limited to 180 W, out of which 60 W is used for recharging the battery drained during eclipse period. A deployable solar panel configuration is also being studied to enhance the power generation constraint experienced presently. In future, it is planned to have a deployable solar panel for more power generation.

Fig. 4 Solar panel configuration

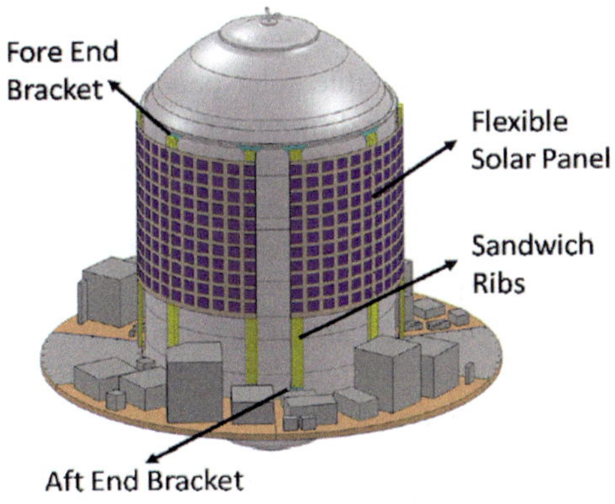

2.4 Telemetry and Tele-Command System

S-band Digital Turbo coded Telemetry transmitter of 0.5 W power output with OQPSK modulation with data rate of 1Mbps is used as exclusive telemetry transmitter for PS4-OP. The Tele-command (T/C) function is realized through S-Band T/C receiver and Telemetry and Tele-command Processing Unit (TTPU) package. S-band T/C Receiver is with a data rate of 4kbps and uses FM/PSK demodulation to get the CCSDS T/C messages. Onboard data storage and playback for housekeeping data and payload data is implemented through TTPU using NAND Flash memory, which can be configured for multiple storages and playback through tele-commands.

3 Results of PS4-OP 1st and 2nd Phase Experiments

3.1 Maintaining the PS4 Stage for 10 Orbits in PSLV-C38

After the separation of all satellites, PS4 stage of PSLV-C38 vehicle was maintained in a 350 km SSO (Sun Synchronous Orbit) orbit for 10 orbits using dedicated Li-ion batteries. PS4 stage was 3 axis stabilized and was rolled so as to point a hypothetical/dummy payload toward earth. The only difference compared to normal mission was that the 3-axis stabilization was carried out by using 50 N bi-propellant thrusters operated in PWPFM mode control scheme.

The health of all propulsion and control system elements were normal throughout the experimental phase. As expected, battery voltage gradually reduced from 29.9 to 27.1 V. Minimum and maximum temperature measured on the battery was 14 and 27 °C.

3.2 Solar Power Generation for PS4-OP in PSLV-C45

The proposed power generation scheme was implemented in PSLV-C45 launch vehicle with flexible solar panel. In addition to the wrap around solar panel, additional power conditioning package was also added on to the Equipment Bay of PSLV-C45 vehicle. Spent stage of PS4 of PSLV-C45 was in 485 km SSO orbit. The power generation system generated 180 W of power when illuminated in the intended orientations. This was conditioned and used to charge onboard batteries which in turn powered the PS4-OP.

PS4-OP-C45 carried onboard 3 scientific payloads namely, 1. AIS (Automatic Identification System) from SAC/ISRO, used for maritime satellite application, where AIS messages from ships are collected and sent back to ground station, 2. APRS (Automatic Packet Repeating System) from AMSAT, India, used to assist amateur radio operators in tracking and monitoring, and 3. ARIS (Advanced

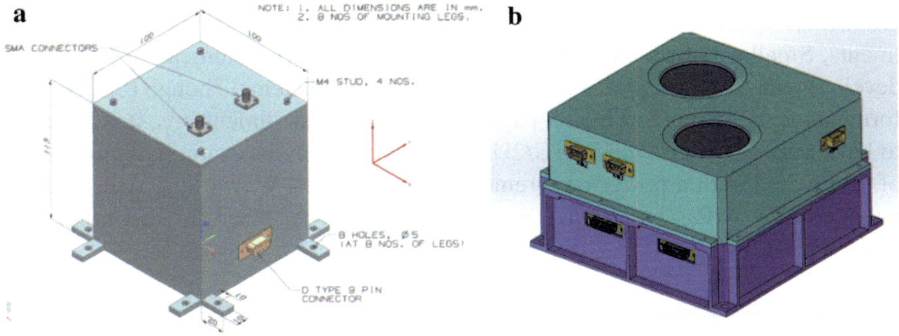

Fig. 5 **a** APRS payload **b** ARIS payload

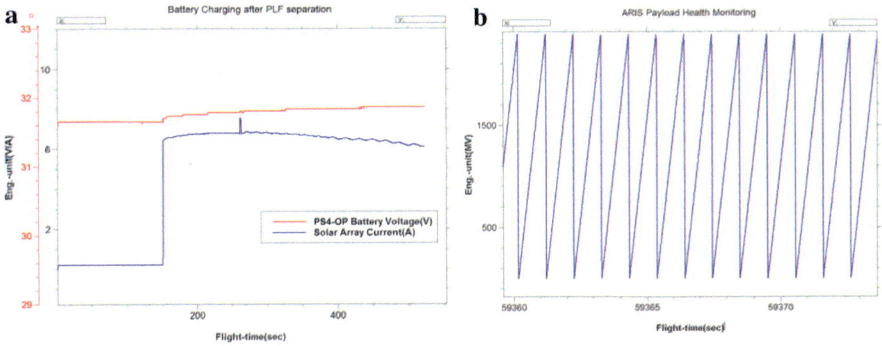

Fig. 6 **a** 1st battery charging on exposure of solar panel **b** ARIS payload health monitoring

Retarding potential analyzer for Ionospheric Studies) from Indian Institute of Space Science and Technology, used for the structural and compositional studies of atmosphere (Fig. 5).

All the three payloads were qualified and acceptance tested to environmental test levels specified for PSLV, which included thermal, thermo-vacuum, vibration and shock tests. Passive thermal management using multi-layer insulator (MLI) and plasmask was employed for this mission. The minimum and maximum ambient temperature observed was in the order of – 18 to + 120 °C. All the three payloads were found to function normally as indicated by the telemetered data (Fig. 6).

4 Conclusion

With PS4-Orbital Platform, ISRO aims to provide scientific community and enthusiastic young students the opportunity to carry out their experiments in space. Payload avenues in space are unending, and ranges from Microgravity experiments, Robotic arm/Smart Space robot technology demonstration, Rendezvous and docking experiments, Small satellites technology development, New communication technology development, Low cost platform for testing Inflatable systems. Though PS4-OP platform can be configured to mount payloads with different dimensions, it is preferred to have dimension in standard 1U/2U/3U sizes [3], with maximum payload weight of 10 kg, and with a power requirement of less than 10 W, working with an input voltage of 28 V. Passive thermal management is only planned for first mission of PS4-OP.

4.1 Announcement of Opportunity (AO) for In-Orbit Scientific Experiments on Orbital Platform (PS4-OP) [4]

Proposals are solicited from National/International Scientific Community for novel space-based experiments to be configured in PS4-Orbital Platform. The AO has a specific objective to invite important payloads for inclusion in the OP mission to strengthen/complement the space-based research activities taking place in industries/academia. It assumes that there will be no exchange of funds under this activity between ISRO and proposing teams.

4.2 Guidelines for Development of Instrument/payload

- Payloads are preferred to be in standard dimensions—1U/2U/3U sizes. Any waiver for mechanical interface is to be obtained from ISRO prior to design finalization.
- Payloads are preferred to have MIL grade components and D-type or circular MIL connectors. Use of commercial/industrial grade components is subject to satisfactory completion of specified acceptance tests.
- Payloads shall be qualified/acceptance tested to the environmental test levels specified for PSLV.
- RF payload frequency/power and sensitivity shall be finalized only after payload interference study with the existing RF elements in PS4-OP/passenger payloads.

References

1. Millan RM, von Steiger R (2019) Small satellites for space science: a COSPAR scientific roadmap. Adv Space Res 64(8)
2. Virgili B, Krag H (2015) Small satellites and the future space debris environment
3. Puig-Sauri J, Twiggs RJ (2005) CUBESAT: design specifications document. California Polytechnic State University and Stanford University's Space Systems Development Laboratory
4. ISRO (2019) Announcement of opportunity (AO) for orbital platform: an avenue for in-orbit scientific experiments. https://www.isro.gov.in/update/15-jun-2019/announcement-of-opportunity-ao-orbital-platform

Sanket—Technology Demonstration of Antenna Deployment System on PSLV Stage 4 Orbital Platform

K. Jagdale, M. Munjal, P. Kurrey, A. Wakode, P. Lohiya, P. Shrivas, A. Sikka, S. Bhansali, A. Kejriwal, A. Vadladi, A. Kumar, A. Savarkar, H. Gidewar, H. Agrawal, M. Dhaka, P. Kasat, R. Shinde, S. Laddha, A. Yadav, A. Mehta, S. Dhanush, I. Phansalkar, J. Saboo, K. Verma, L. Chaudhari, N. Debnath, S. Athaley, S. Sabnis, V. Verma, V. Gala, and Y. Jindal

Abstract An Antenna Deployment System has become an essential component of any pico- or nano-satellite design due to space constraints during launch. The Sanket mission is a technology demonstration designed to be flown on the Indian Space Research Organization's PSLV Stage 4 Orbital Platform (PS4-OP) (Announcement of opportunity for orbital platform. Indian Space Research Organisation, Bangalore [1]) and aims to qualify the team's Antenna Deployment System in Ultra High Frequency band to a TRL-7 (Technology Readiness Level) in Low Earth Orbit (LEO). Sanket, i.e. the complete system, comprises an ADS and an auxiliary system. The purpose of the auxiliary system is to test the ADS on PS4-OP simulating a 1U CubeSat mission life cycle and conditions. Sanket will be mounted on PS4-OP which remains in LEO for around 6 months. Our Antenna Deployment System is developed as an independent module that is compatible with standard CubeSat sizes 1U, 2U and 3U.

Keywords Antenna deployment system · CubeSat · PS4-orbital platform · Student satellite · Technology demonstration · UHF antenna

Abbreviations

ADS	Antenna Deployment System
AUX	Auxiliary System
Downlink	Signal transmitted from deployed antenna to ground station
EEPROM	Electrically Erasable Programmable Read-Only Memory

K. Jagdale · M. Munjal · P. Kurrey · A. Wakode · P. Lohiya · P. Shrivas · A. Sikka · S. Bhansali · A. Kejriwal · A. Vadladi · A. Kumar · A. Savarkar · H. Gidewar · H. Agrawal · M. Dhaka · P. Kasat · R. Shinde (✉) · S. Laddha · A. Yadav · A. Mehta · S. Dhanush · I. Phansalkar · J. Saboo · K. Verma · L. Chaudhari · N. Debnath · S. Athaley · S. Sabnis · V. Verma · V. Gala · Y. Jindal
Student Satellite Lab, Aerospace Department, IIT Bombay, Mumbai 400076, India
e-mail: ritulshinde@gmail.com; iitbssp@iitb.ac.in

© The Author(s), under exclusive license to Springer Nature Singapore Pte Ltd. 2023
R.S. Priyadarshini and T. Sundararajan (eds.), *Advances in Small Satellite Technologies*, Lecture Notes in Mechanical Engineering, https://doi.org/10.1007/978-981-19-7474-8_8

HM	Health Monitoring
PCB	Printed Circuit Board
Telemetry	Signal transmitted from the antenna on PS4-OP
Telecommand	Signal transmitted to the antenna of PS4-OP
Uplink	Signal transmitted from ground station to the deployed antenna

1 Introduction

"Sanket", literally meaning "Signal" in Sanskrit, is the name of the mission to demonstrate the ADS, designed and developed by IIT Bombay Student Satellite Program for CubeSat applications. As the number of CubeSat missions in India continues to grow, the need for indigenously developed, reliable communication systems become more pressing. Various international firms manufacture and sell Antenna Deployment Systems for CubeSat applications. However, these are associated with steep costs and accessibility issues. The reliable power and telemetry subsystems of PS4-OP provide the perfect opportunity to test the ADS. The payload is verified by establishing a half-duplex communication link between the deployed antenna and the ground station. Post successful technology demonstration, technology will be transferred to the Indian industry to aid and promote future CubeSat missions in India as an indigenous solution.

2 System Description and Methodology

2.1 Mechanical Structure

The ADS (Fig. 1) comprises an in-house manufactured antenna, support and interface structures for the antenna, a deployment mechanism and a deployment detection circuit. The antenna, made of stainless-steel tape spring, is rolled and held inside the module with the help of a flexible PVC/PVA sheet. The retention and release mechanism uses a nichrome burn-wire design that releases the stowed antenna by thermally cutting a nylon thread held in tension connected to a flexible PVC sheet. Deployment is detected using Single Pole Double Throw (SPDT) switch, one for each pole of the dipole antenna. The ADS will be mounted on the AUX whose structure is inspired by Advitiy (2nd Student Satellite of IIT Bombay) a standard 1U CubeSat [2]. The structure is made by Al-6061 and the PCB is made of FR04.

The system design is carried out through an iterative process from making the configuration layout to its CAD, simulating it and then manufacturing it for testing and checking the manufacturability to resolve the concerns in the next iteration. Prototypes and engineering models are manufactured for better visualisation and testing

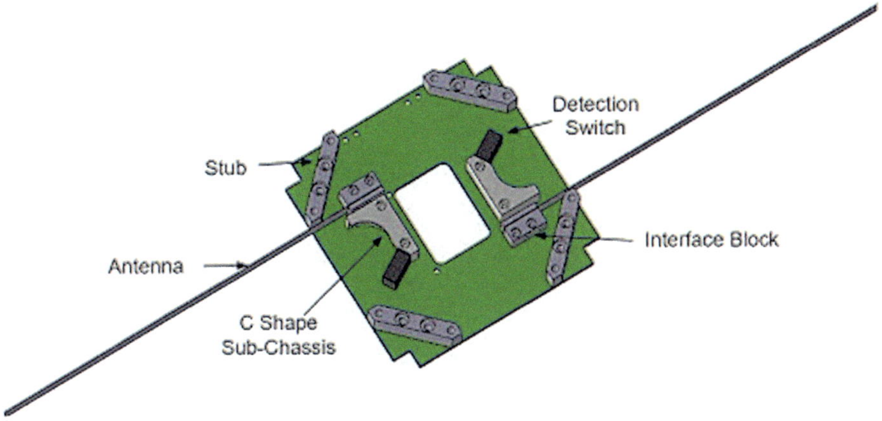

Fig. 1 ADS

Table 1 System description

	Dimensions	Mass (g)
ADS	98 mm × 98 mm × 8.6 mm	74.90
Sanket (ADS + AUX)	126 mm × 433 mm × 100 mm	713.05

purposes whereas Qualification Model and Flight Model are made for complete rigorous testing and launch, respectively.

Static, harmonic, modal and random vibrations' simulations are performed for PSLV launch loads on the structure, and stress analysis is carried out. These, as well as similar simulations for thermal analysis [3], are performed in ANSYS. These simulations are done to ensure that the system is safe to fly on the PSLV and faces no damage during the flight and later in orbit. The system mechanically integrates to PS4-Orbital Platform through 8 M6 screws via the protrusions present on the lower chassis (Fig. 2; Table 1).

2.2 *Electrical Design*

Sanket accommodates AUX, Communication and ADS PCBs (refer Fig. 2). The system will electrically interface with the PS4-OP via D-connector to receive unregulated power of 28 V-10 W. The fluctuations will be mitigated by the EMI filter (SVRMC28) and the voltage regulator (SVRHF283R3S) will step down the voltage to a fitting 3.3 V to power the whole system. Current limiters (TPS7H2201-SP) are used to avoid overcurrent damage of the components on ADS and Communication PCBs and also act as controlled switches. The AUX PCB has a microcontroller called

Fig. 2 Sanket

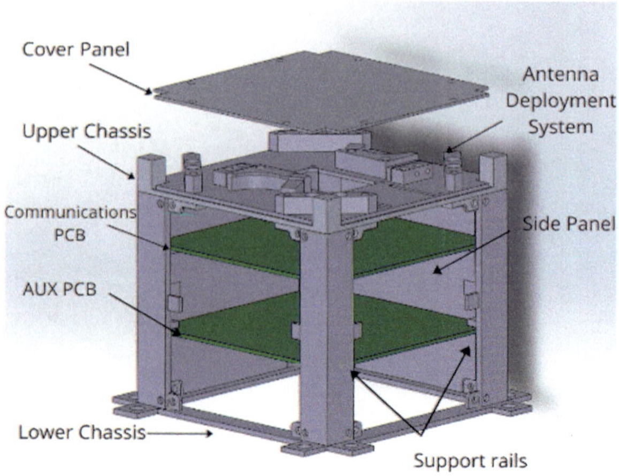

the AUX microcontroller (ATmegaS128) which is responsible for the scheduling of tasks, communicating with the PS4-OP for telemetry and telecommand, collecting HM data and enabling the power supply to the Communication and ADS PCB. The collected HM data is stored on the EEPROM (AT69170F). When scheduled, the ADS PCB would use the buck converter (TPS-50601A-SP) to regulate the voltage received from AUX PCB and heat the nichrome wire which would thermally cut the nylon wire to deploy the antenna. All the PCBs are designed on EAGLE with dimensions 92 mm × 92 mm. The flight codes are being written using Atmel Studio and are simulated on Proteus. Version control is done using Git.

2.3 Communication Design

A single deployable dipole antenna designed to work in UHF band (435–438 MHz) will be used to communicate with the ground station. The UHF band requires small antenna pole length and hence is ideal for CubeSat use. Antenna is designed and simulated in HFSS software and optimized for minimum S11 in the required frequency band. Tapered traces on ADS PCB ensure impedance matching between the antenna and the feed. The Communication PCB is responsible for RF communication. A single transceiver (CC1125), programmed by the communication microcontroller (ATmegaS128), will handle both the uplink as well as the downlink channel. The communication microcontroller will also collect the HM data from AUX microcontroller which will be transmitted along with the identity (name + callsign) as OOK-modulated downlink. A high-power amplifier (CMX901) will amplify the downlink signal before it is transmitted by the UHF dipole antenna. The GFSK-modulated uplink commands received by the antenna will be amplified by a low-noise amplifier

(MAAM-011229) before being demodulated. The switching of signals between the uplink and downlink channels will be done by a high-power switch (HMC574A).

The design development will proceed in the following order:

Prototype > Engineering Model > Qualification Model (QM) > Flight Model (FM)

A prototype is made to prove the concept, engineering model is a working system with cost components; qualification model and flight model will be made up using the space-grade components. The design process followed by the team is based on the V-model derived from the principles of Systems Engineering. All requirements on every subsystem are listed, and a checklist or a set of test cases to qualify the requirements is prepared to use at each development stage for Quality Assurance (QA). An Interface Control Document (ICD) is maintained which defines how all parts integrate into the final design. Failure Mode and Effect Analysis (FMEA) is performed for the system to eliminate and reduce failures.

3 Analysis

Electrical components need to be shielded from ionizing radiation. For this, simulations were performed on SPENVIS and it was decided to cover the system with panels of 3-mm thickness to effectively reduce the effects on crucial electronics. Fixtures are designed to facilitate the integration of SANKET and ADS.

Bolt preload study is done to find fasteners and the optimal preload (torque) that should be applied to screws, nuts and bolts to reduce the chances of failure at all the joints. The AUX and communications PCBs are fixed, one from above and one from below (refer Fig. 2) to tackle integration constraints.

All the electrical components selected are of space-grade qualification and radiation-hardened or having space heritage with an operating temperature range of −55 to 125 °C, selected after considering power requirements and efficiency parameters in accordance to the power budget. All on-board decisions will be made by the AUX microcontroller, including switching between uplink and downlink. High power switch will be used to facilitate this switching between uplink and downlink channels instead of a circulator because the circulator has larger dimensions and more weight. Health monitoring data will be collected and analysed on-board at regular intervals so that any unusual voltage or current values can be detected immediately. Apart from this, all the data will be stored and sent through PS4 telemetry for future analysis. HM data received through downlink will be cross-checked with the data received through telemetry to ascertain the functioning of the deployed antenna. OOK modulation has been chosen for downlink since it has the best signal to noise ratio (least required threshold) out of all the modulations. GFSK will be used for uplink modulation as it provides higher data rates with a better noise immunity and moderate signal to noise ratio requirement. The uplink and the telecommand can

Table 2 Structural simulation results

	Static (Max. stress in MPa)	1st Frequency (Hz)	Harmonic (Max. stress in MPa)	Random vibration (Max stress MPa)		
				x-axis	*y*-axis	*z*-axis
Components	Bottom cage	PCB	Support rail	Bottom cage	Bottom cage	PCB
Operational limits	275	90 (min.)	275	275	275	120
Simulation results	11.49	252.76	5.22	29.45	32.18	54.63
FOS	23.93	–	52.68	9.34	8.55	2.20

be used to disable or restart the Sanket. Also, provision for an uplink command for verification of uplink has also been made. Acknowledgement of the instructions uplinked will be taken in telemetry to test uplink.

The team has a dedicated Ham team and a ground station segment which has developed expertise in satellite tracking through regular satellite tracking sessions. The team receives and processes downlink data received from the satellites.

4 Results and Discussion

4.1 *Structural Simulations*

In order to investigate the structural safety of Sanket during the launch, equivalent (Von-Mises) (denoted by *E*) stresses from various types of simulations are checked. The factor of safety (FOS) is calculated for every part of the system from ANSYS simulation results (Table 2).

Thermal Simulations Result—Simulated a simplified model of Sanket on ANSYS for one orbit period. Over the time period, the maximum temperature of 79.362 °C was recorded on the C-Shaped sub-chassis and the minimum temperature of − 85.402 °C was recorded on the antenna. High-power amplifier, EMI filter and voltage regulator are the most critical components in the system which radiated maximum thermal radiation.

4.2 *Deployment Tests*

These tests were done on the acrylic prototype of ADS (refer Fig. 3a, b) in order to find the range of current across the nichrome wire that would ensure the cutting of nylon wire in around two seconds. A digital power source was used to generate a potential difference between the ends of nichrome wire which is 32 AWG and 1.5 cm

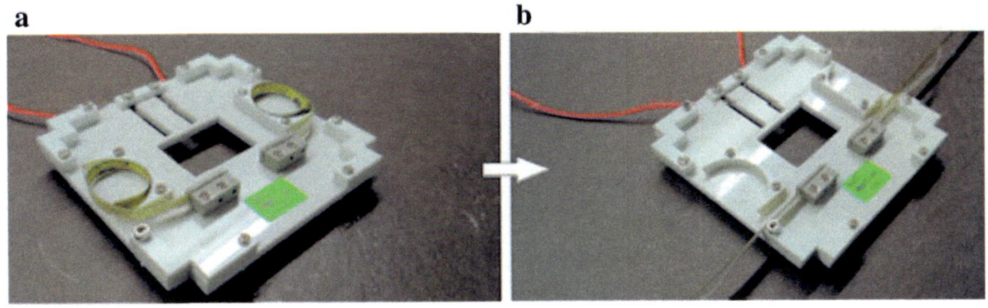

Fig. 3 **a** Before deployment **b** Post deployment

in length, and the test was concluded under 4 V. When tested in room temperature, it is observed that the time to break the nylon thread is less than 2 s when the current in nichrome is around 1.2 A. At this current, the temperature of nichrome reaches more than 200 °C which is significantly higher than the temperature of PS4-OP during launch. This ensures that thermal cutting is not initiated automatically, due to unexpected temperature variations of the system. Double-stranded braided nylon was tested which showed improved strength and similar thermal cutting time as compared to its single-stranded counterpart. A better alternative for nylon, Vectran, will be used for further tests.

4.3 Antenna Simulation

Modelling of antenna was done in HFSS-15 by assuming the PS4-OP as a hollow aluminium cylinder of radius 1 m, length 2 m and thickness 0.2 m. The CubeSat was modelled as a 1U hollow aluminium box and placed on one of PS4-OP's flat surfaces. The dipole antenna's poles are modelled as stainless-steel strips with width 6 mm, thickness 0.1 mm and variable length (to be optimized) (Figs. 4, 5 and Table 3).

One pole of the dipole antenna is grounded and the other pole is the feed. The PS4-OP is taken as the ground. It was observed that in all the cases, the radiation pattern is asymmetrically skewed towards the direction of the feed pole. The most favourable S11 (lowest) and radiation pattern (high gain and high beamwidth) were obtained for pole length = 184 mm.

5 Conclusion

Antenna simulations show that the radiation pattern gets highly distorted due to the large metallic body of PS4-OP. The gain of the antenna is significantly higher than the conventional dipole antennas making the radiation pattern directive. This puts

Fig. 4 3D radiation pattern

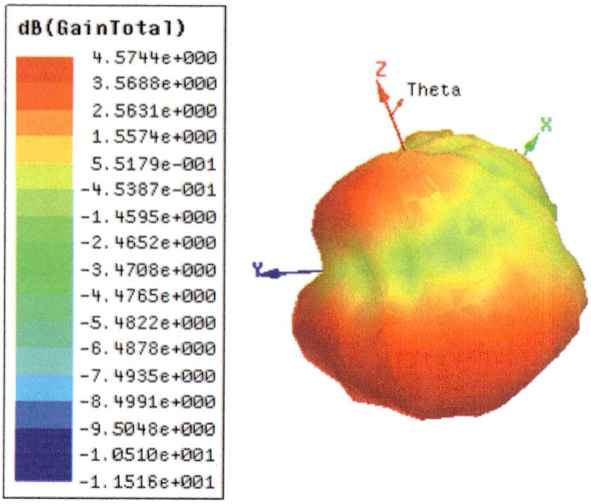

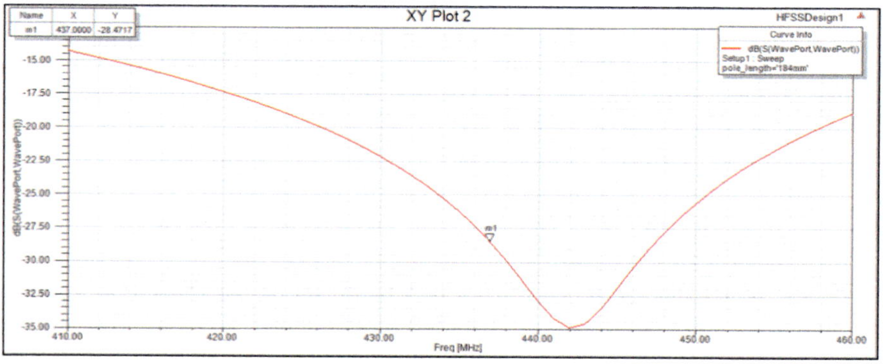

Fig. 5 S11 plot

Table 3 Simulation for pole length

Pole length (mm)	S11 at 437 MHz (dB)	Freq of min S11 (MHz)	Min S11 (dB)	Max gain (dB)	HPBW (degrees)
180	− 18.83	456	− 31.82	4.45	75.46
182	− 23.38	447	− 40.29	4.47	73.53
184	− 28.47	442	− 34.85	4.57	72.47
186	− 25.41	432	− 26.66	4.49	75.24

the requirement of precise control on the platform for continuous signal reception. Due to the unavailability of anechoic chambers, it is decided to measure the radiation pattern in open ground to minimize the inaccuracy due to signal absorption.

The deployment test shows that the deployment time is less than 1 s at room temperature, but it increases as the surrounding temperature decreases. This encourages the necessity of deployment testing at different temperature and pressure conditions. Following this need, manufacturing of vacuum chamber has been started. The structural simulations indicate that every mechanical component has FOS > 1.5. The mission is at a stage where the design is finalized and component level testing has started according to the V-model design approach.

The reliable power and telemetry subsystems of PS4-OP provide the perfect opportunity to test the ADS for different communication modes. The telemetry will help to closely monitor the system status even if the deployment is failed. This gathered data of system status will be crucial for future missions.

Acknowledgements We would first like to express our sincere gratitude to the whole of the ISRO team for inspiring young minds to venture into the domain of space science and technology and guiding us over the last 12 years. We would thank Prof. Varun Bhalerao, Prof. Prabhu Ramachandran and Prof. Arnab Maity, our Faculty Advisors for giving us timely inputs and suggestions as also freedom and space for our design. We extend thanks to Industrial Research and Consultancy Centre (IRCC), IIT Bombay for providing us with the funds needed for a project of this magnitude and the Department of Aerospace Engineering, IIT Bombay for providing us laboratory space. We would like to thank IIT Bombay as a whole for creating an environment conducive for growth and excellence. Finally, we would like to thank the families of our team members who support us.

Authors and coauthors: Karan Jagdale, Mrigi Munjal, Prashant Kurrey, Ashay Wakode, Pushkar Lohiya, Puneet Shrivas, Anmol Sikka, Sanskriti Bhansali, Abhishek Kejriwal, Amrutha Lakshmi Vadladi, Ankit Kumar, Atharv Savarkar, Hemant Dilip Gidewar, Hrithik Agrawal, Mohit Dhaka, Pranav Kasat, Ritul Shinde, Shreya Laddha, Adesh Yadav, Akshat Mehta, S. Dhanush, Ishan Phansalkar, Jayant Saboo, Kriti Verma, Leena Chaudhari, Navjit Debnath, Shreeya Athaley, Shreyas Sabnis, Vidushi Verma, Vineet Gala, Yatin Jindal.

References

1. Announcement of opportunity for orbital platform. Indian Space Research Organisation, Bangalore, 560 231, 2019
2. Kothari H, Lohiya P, Zubair A, Singh S (2020) Satellite structure of Advitiy (Second student satellite of IIT Bombay). In: Advances in small satellite technologies. Springer, pp 535–544
3. Garzon MM (2012) Development and analysis of the thermal design for the OSIRIS-3U CubeSat

Development of a PS4-OP Payload for Technology Demonstration of Small-Satellite Subsystems

T. K. Anant Kumar, Ch. Sai Abhishek, Vivek Garg, Yugal Joshi, Anantha Datta Dhruva, Mallikarjun Kompella, Ishan Sarvaiya, S. V. Janakiram, E. Harshavardhan, Sandeep Prasad Shaw, Divyansh Prakash, Devashish Bhalla, Sankalp Vishnoi, Joji John Varghese, P. Suhail, H. Priyadarshan, M. S. Harsha Simha, V. S. Sooraj, and P. Raveendranath

Abstract Following the Indian Space Research Organisation's (ISRO) novel idea of using the spent PS4 stage of the PSLV as a platform for scientific experiments and technology demonstrations, a unique opportunity has been created to develop and test small-satellite subsystems and instruments as PS4-OP (PSLV 4th Stage Orbital Platform) payloads, prior to the launch of the small satellite itself. The PiLOT (PS4 in-orbital OBC and TTC) payload has been developed with an engineering objective of flight qualifying the in-house developed small-satellite subsystems, which are the OBC (On-Board Computer) and TTC (Telemetry and Telecommand) subsystems. The PiLOT payload also contains a RADFET (Radiation Field Effect Transistor) sensor for monitoring radiation dosage. Mapping the temporal and spatial distribution of the radiation dosage will help improve the understanding of the space weather in the low earth orbit. This paper describes in detail the development of the PiLOT PS4-OP payload including a description of the different subsystems of the payload, which are the RADFET sensor board, the OBC, the TTC board, and the PS4-OP interface board. The paper concludes with a description of the integration, testing methodology, and the current developmental status of the payload.

Keywords PS4-OP · Radiation dosimetry · RADFET · On-Board computer · Flight software

T. K. Anant Kumar (✉) · Ch. S. Abhishek · V. Garg · Y. Joshi · A. D. Dhruva · M. Kompella · I. Sarvaiya · S. V. Janakiram · E. Harshavardhan · S. P. Shaw · D. Prakash · D. Bhalla · S. Vishnoi · J. J. Varghese · P. Suhail · H. Priyadarshan · M. S. Harsha Simha · V. S. Sooraj · P. Raveendranath
Indian Institute of Space Science and Technology, Thiruvananthapuram 695547, India
e-mail: anant.telikicherla@ualberta.ca

H. Priyadarshan
e-mail: priyadarshnam@iist.ac.in

© The Author(s), under exclusive license to Springer Nature Singapore Pte Ltd. 2023
R.S. Priyadarshini and T. Sundararajan (eds.), *Advances in Small Satellite Technologies*, Lecture Notes in Mechanical Engineering, https://doi.org/10.1007/978-981-19-7474-8_9

1 Introduction

Recently, the Indian Space Research Organisation (ISRO) released an announcement of opportunity inviting proposals to develop payloads that can be tested on the PS4-Orbital Platform (PS4-OP) [1]. This platform is a novel idea formulated by ISRO to use the spent fourth/final stage of the Polar Satellite Launch Vehicle (PSLV)), called the PS4, to conduct in-orbit scientific experiments. Conventionally, the final stage of typical launch vehicles is rendered useless after the completion of the mission, acting as space debris till their re-entry burn. By re-purposing the spent PS4 stage as an orbital platform that can provide a payload with power, a communication interface, attitude stabilization, and control, ISRO has provided a unique opportunity to develop payloads that can perform in-orbit experiments, while remaining fixed to the PS4-OP [2]. Previously, three PS4-OP payloads have been tested in the orbital platform of the PSLV-C45 (launched in April 2019) which includes the Advanced Retarding Potential Analyzer for Ionospheric Studies (ARIS) developed by IIST [3]. According to the Announcement of Opportunity (AO) released by the ISRO, the mass of PS4-OP payloads must be less than 10 kg, with a volume of 1U/2U/3U and power consumption of less than 10 W. These constraints are especially suitable for small-satellite subsystems as their mass, power, and volume requirements lie in these ranges. Thus, using the PS4-OP, critical small-satellite subsystems, as well as science payloads, can be tested prior to the small-satellite launch which will help characterize and qualify these subsystems. The PiLOT mission is a PS4-OP payload developed indigenously by the Small-spacecraft Systems and PAyload CEntre (SSPACE) at IIST. The PiLOT mission aims to space qualify the in-house-built small-satellite subsystems, namely the On-Board Computer (OBC) and the Telemetry and Telecommand subsystem (TTC) subsystems. In addition to this, the PiLOT payload also contains a low-cost sensor for in-situ radiation dose monitoring. These subsystems are slated to fly on IIST small-satellite missions (such as the Ahan mission) in the coming years. The key specifications of the PiLOT payload are specified in Table 1, which are described in detail in this paper.

2 PiLOT Payload Design

The PiLOT mission is developed in a 1U CubeSat form factor. The payload consists of four boards, the On-Board Computer (OBC), the Telemetry and Telecommand (TTC) Board, the RADFET board, and the PS4-OP interface board, which are connected to each other using the PC104 standard cubesat bus. The PiLOT payload requires a single 5 V supply input from the PS4-OP which is then converted to the different voltage levels internally, on the PS4-OP interface board. For sending telemetry data to the PS4-OP bus, an RS485 interface is provided. The TTC board has an RF output (SMA connector) that can act as a redundant mode of data transfer in addition to the

Table 1 PS4-OP orbital requirements

Dimensions and mass	130 mm × 100 mm × 100 mm (approx.), 1 kg (approx.)
Power	2 W (Nominal), 7 W (during RF transmission)
Science objective	To conduct in-situ radiation dosimetry in the LEO (low earth orbit) using a power-efficient low-cost RADFET based sensor
Technology demonstration objective	To flight qualify functionalities of in-house-designed OBC and TTC
Science instrument specifications	Radiation dosage—0 to 100 KRad, Resolution—1 Rad
OBC (On-Board Computer) board specifications	Microsemi SmartFusion2 SoC FPGA, 8 Kb Flash Memory, 1 × 128 Gb SD Cards
TTC (Telemetry and Telecommand) board specifications	RF output power: 0.5/1 W, telecommand sensitivity: −120 dBm, frequency of operation: 145.825 MHz

telemetry channel of the PS4-OP. Figure 1 shows the electrical block diagram of the payload, including power and data signals. A detailed description of the subsystems is given in this section.

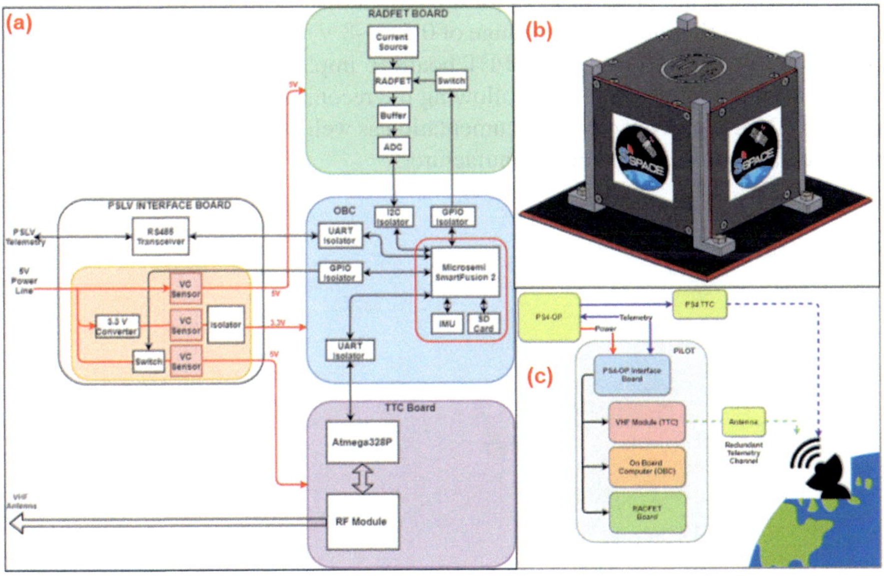

Fig. 1 **a** PiLOT system block diagram, **b** PiLOT outer structure, **c** PiLOT mission concept

2.1 RADFET Board

The PiLOT mission utilizes low-cost technology to conduct the radiation dosage measurement using a RADFET (Radiation Field Effect Transistor). This is a discrete p-channel MOSFET that is optimized for radiation sensitivity and can detect ionizing radiation such as gamma rays, x-rays, electrons, and high-energy protons. RADFETs have been previously used by organizations such as the European Space Agency as radiation dosimeters on various missions such as Nanosat-1B [4]. Through generation and trapping of radiation-induced charges in the gate oxide, radiation exposure changes the output voltage of the RADFET. The VT02 RADFET developed by Varadis is used for this mission and can measure a total accumulated dose from 0 to 100 Krad [5]. The RADFET response is non-linear and a pre-recorded calibration curve is used to read the dose. Figure 2b shows a calibration curve for a range of 0–100 kRad. The RADFET operates in two modes namely irradiation and read-out. In the irradiation mode, the RADFET measures radiation, and all terminals of the RADFET are grounded. In the read-out mode, the current is applied to the RADFET and the voltage measurement is taken. A PCB has been designed in house (Fig. 2a) in order to implement the read-out circuit for the RADFET using a high-resolution ADC. Since RADFETs can measure doses between 1 cGy (1 rad) and 1 kGy (100 krad), using this ADC and low-noise circuit design and components, the board aims to provide a resolution of 1 Rad. The read-out is done by forcing a DC current (in the order of μAs for a fixed amount of time) into the device and measuring the DC voltage (in the range of 0.5 to ~8 V). A current source, switch and ADC is implemented on the RADFET board to implement this read-out circuitry. The read-out circuit is created by following the recommended guidelines according to the RADFET read-out circuit documentation as well as temperature compensation documentation provided by the manufacturer.

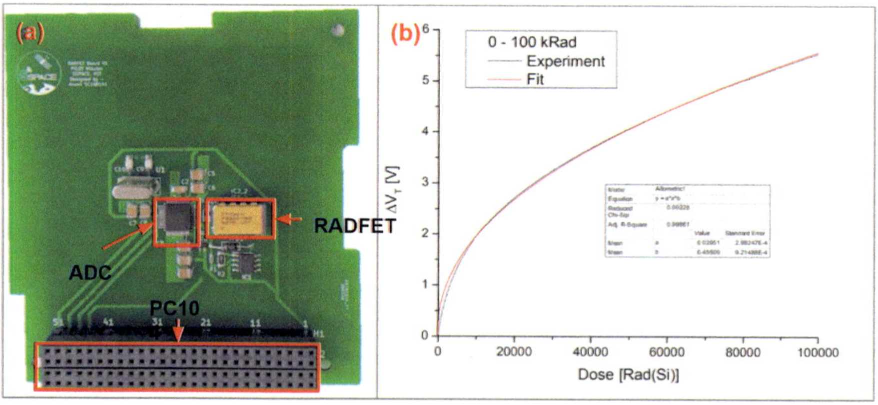

Fig. 2 **a** RADFET board EM, **b** RADFET calibration curve (0–100 kRad)

2.2 *On-Board Computer*

The On-Board Computer is built around the SmartFusion2 M2S090 SoC (System on Chip) FPGA which incorporates ARM Cortex M3 microprocessor, SRAM (Static RAM), eNVM (electronic non-volatile memory), RTC (real-time counter), and various other features. Since the Microsemi SmartFusion2 SoC FPGAs and IGLOO2 FPGAs are 65 nm, non-volatile, and flash-based FPGAs, they exhibit intrinsic robustness against radiation-induced single event upsets. This is in direct contrast to SRAM-based which exhibits single event upsets in the SRAM configuration memory that configures the function of logic cells and connects routing tracks together. This can lead to unpredictable behavior that can be very harmful to the small-satellite system. In order to demonstrate absence in configuration upsets due to radiation, Microsemi has conducted radiation testing on its M2S050 SmartFusion2 FPGAs. Forty-eight units of Microsemi M2S050 SmartFusion2 FPGAs were exposed to heavy ions at LET (linear energy transfer) levels up to 90.3 meV-cm^2/mg, and no configuration upsets were detected in testing on a total of 48 parts in a total influence of 2.83×10^9 heavy ions [6]. The OBC board requires a 3.3 V supply line to work and also includes one SD card slot, 64 Mb SPI-Flash, an ADC, an external processor supervisor circuit, and a 9-axis IMU (Inertial Measurement Unit). In order to make the SoC isolated from the other subsystems of the satellite, power and signal isolators have been used between the SoC and the other subsystems. Figure 3a shows the engineering model of the OBC developed, which is currently undergoing comprehensive performance tests to ensure robust functionality. Once the OBC's functionality is verified, the IMU on the OBC will also undergo calibration. For the PiLOT mission, the IMU is used just to collect accelerometer, gyroscope, and magnetometer values for housekeeping purposes. If similar telemetry values are also available from sensors on the PS4-OP, then the measurements of the OBC's IMU can be validated by comparison during post-processing. The OBC is responsible for collecting housekeeping (HK) data from all subsystems and science data from the payloads. It is also responsible for running the flight software which will be described in detail in the next section. This OBC is an improved version of the previously designed OBCs developed at IIST for the INSPIRESat-1 and INSPIRESat-2 missions. These previous versions have successfully qualified TRL-8 status by completing the environmental tests (vibration and thermal-vacuum) as a part of the integrated satellites [7, 8].

2.3 *Telemetry and Telecommand (TTC) Board*

The TTC board aims to provide a redundant scheme for downlinking telemetry apart from the PS4-OP's telemetry. This board uses the low-cost VHF transceiver IC with an operating frequency range from 134 to 174 MHz. The board has configurable RF output power which can be set to 0.5 or 1 W, a data rate of 1.2 Kbps, and a Bit Error Rate (BER) of 10–5. Currently, for the PiLOT mission, only a telemetry

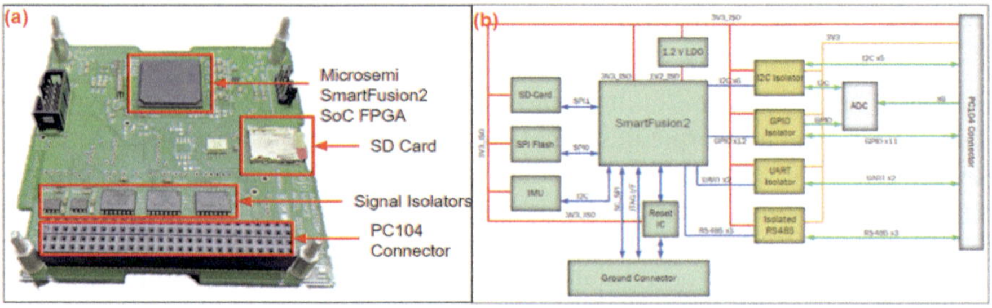

Fig. 3 PiLOT OBC: **a** engineering model, **b** OBC block diagram

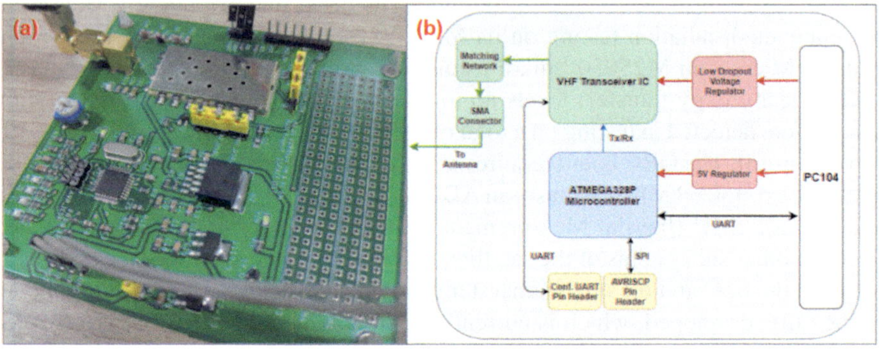

Fig. 4 PiLOT TTC: **a** engineering model, **b** TTC block diagram

downlink is planned, but if required the TTC board also has telecommand reception capability with a sensitivity of −120 dBm. The board also contains a microcontroller that implements AX.25 framing and deframing and implements a UART (Universal Asynchronous Receiver Transmitter) communication protocol. The OBC can use these UART lines to send telemetry packets to the module as required (Fig. 4).

2.4 PS4-OP Interface Board

This board acts as an interface between the PS4-OP and the PiLOT payload. An RS485 telemetry interface is implemented, which is used to transfer data from the PiLOT payload to the PS4-OP's data bus. These lines are passed through Opto-Isolators to ensure electrical isolation between the payload and the PS4-OP's telemetry bus. The board also contains a power interface with the PS4-OP and requires a single 5 V line to operate. As shown in the block diagram in Fig. 5b, the 5 V supply is used to directly power the RADFET board and the TTC board. The PS4-OP interface board also contains a buck converter followed by an isolation circuit to provide 3.3 V for the OBC board. Lastly, the board also contains a subsystem that enables switches

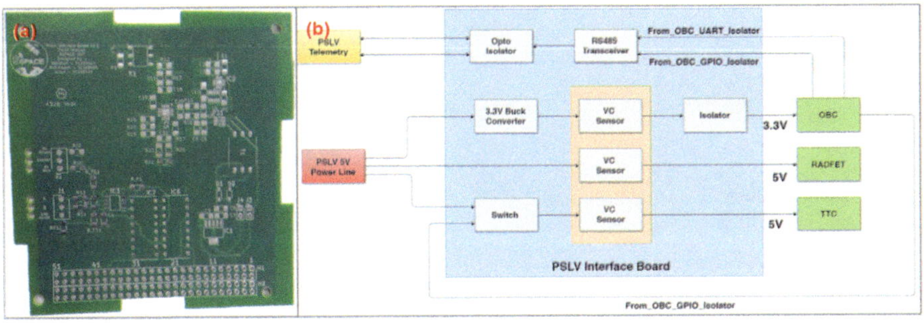

Fig. 5 PiLOT PS4-OP Interface Board: **a** engineering model, **b** block diagram

and voltage-current sensors for the health monitoring of different subsystems. This board is planned to go through official reviews by the PSLV team to ensure that the power and data interfaces are compatible. Figure 5a shows the engineering model PCB of the PS4-OP adapter which is planned to be assembled and tested at IIST in the coming months.

3 Flight Software and Payload Operations

The flight software for the PiLOT payload is written in C as a sequential program. The C code runs on the ARM CortexM3 Microcontroller implemented in the Microsemi SmartFusion2 SoC (System on Chip) FPGA. The FPGA implements and interfaces with the RADFET board and the PS4-OP interface board. In addition to the peripherals for interfacing, custom-developed Verilog modules are also implemented in the FPGA, including a Watchdog Timer Handler, a Real-Time Counter, and a Launch Vehicle Telemetry Interface Handler. This section describes the sequence of operations of the flight software that is illustrated in the flowchart in Fig. 6.

When power is received from the PS4-OP, the OBC and the RADFET board are turned ON automatically. The power to the TTC board is controlled by the OBC using a switch, which is turned ON after completing the initialization procedure of the OBC. After power ON of the OBC, the initialization procedure takes place which includes initialization of the various memories, peripheral drivers, software parameters, and variables. After this, the program enters into an infinite while loop where different tasks are carried out sequentially by the scheduler. The flight software of the PiLOT mission doesn't have an operating system and also doesn't use any interrupts. The scheduler is implemented as a simple sequential C program, and the timing of the tasks is maintained using the Real-Time Counter (RTC) implemented in the FPGA. The first task executed by the scheduler is handling the external watchdog peripheral circuit, and this is done by "petting" the watchdog timer with a toggle signal every one second. This watchdog timer generates a reset to the OBC in case the flight software is stuck in some software routing. Following this housekeeping data from

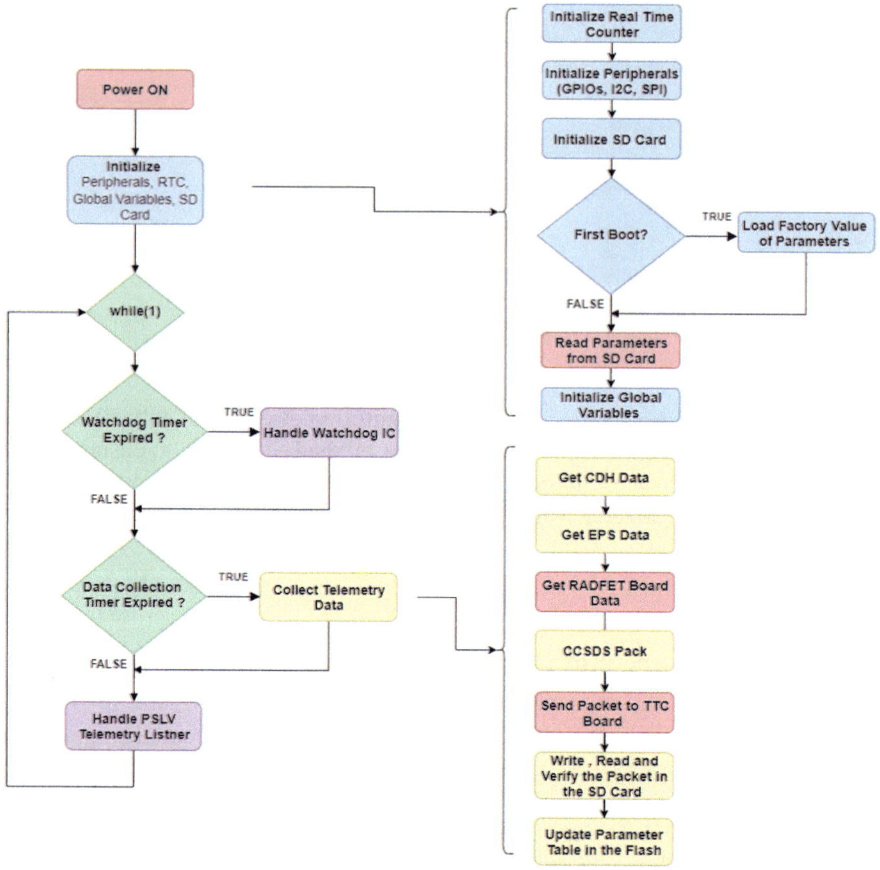

Fig. 6 PiLOT flight software flowchart

the OBC (On-Board Computer), the PS4-OP interface board, science data from the RADFET sensor board is collected using different peripheral interfaces. This data is then converted into a beacon packet following the CCSDS (Consultative Committee for Space Data Systems) packet format. Then, the data packet is transferred to the TTC board to transmit using the redundant VHF telemetry channel. The data packet is then stored in the non-volatile memory (SD Card/ Flash memory) of the OBC, and it is also read back from the memory to verify that it is working properly. Once the data packet is stored, it is copied to the launch vehicle telemetry handler Verilog module implemented in the FPGA. This module then sends the data packet to the PS4-OP telemetry channel following appropriate timing constraints. As mentioned previously, the OBC also has an external watchdog timer IC that triggers a power reset of the SoC in case of any anomalies. When such a reset happens, the continuity in value of certain flight software variables (such as Real-Time Counter value, number of boot-ups, etc.) needs to be maintained. This is done by regularly updating the RTC value in the non-volatile memory of the OBC and reading back the values from the non-volatile memory during the initialization procedure.

4 Testing and Integration

Since all of the subsystems of the PiLOT payload are developed in-house, a rigorous testing mechanism has been adopted to ensure the reliable functionality of all subsystems. This involves first testing the functionality of individual subsystems through comprehensive performance tests (CPTs). In order to conduct the CPT of the OBC and the FSW, the complete code for the sequence of operations was implemented and tested on the OBC. Similar tests were also conducted on the TTC board by sending beacon packets from the module and receiving them in the ground station at IIST. Similar tests are also being conducted for the RADFET board. Once the tests on the individual boards (called Unit Tests) are completed, interface tests are carried out between the different boards. Figure 7 shows the test setups of the OBC and RADFET unit tests as well as the interfacing tests between OBC-TTC and OBC-RADFET. Once the interfacing tests are completed, the boards are integrated together on a lab bench, before integrating them into the mechanical structure.

Figure 8 shows three plots from a 12-h (approx.) long CPT conducted on the OBC. During this test, the entire sequence of operations was run and a test packet was generated every 5 s that was collected in a check-out computer. In order to test the reset mechanism of the OBC, a special function was written which halts the flight software every 30 min (Fig. 8c), to trigger an "artificial" reset from the external processor supervisor circuit. The ability of the code to load the parameters correctly from the non-volatile memory (SD Card) after Power ON was tested. As shown in

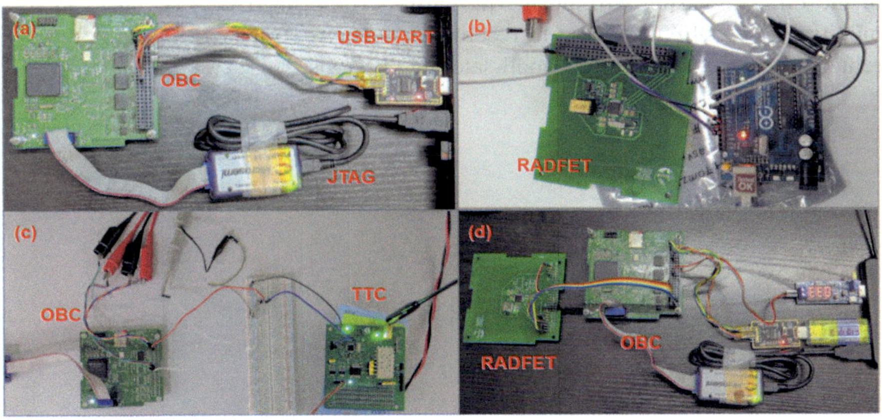

Fig. 7 PiLOT Testing: **a** OBC unit test, **b** RADFET unit test, **c** OBC-TTC interfacing, **d** OBC-RADFET interfacing

Fig. 8a, the value of the Real-Time Counter is increasing continuously throughout the test. This is because the variable parameters (such as the RTC value, reset the counter, etc.) are regularly stored in the non-volatile memory (SD Card). Once a reset happens, the value of the Real-Time Counter is read back from the SD card and updated to ensure that the continuity is maintained. In order to verify the functionality of the SD Card, a code was implemented on the SoC to successively write a unique data array to each of the sectors of the SD Card. The array is then read back and compared to the original array to check for any errors. If both the arrays are the same, the write-read-verify counter variable is incremented indicating that iteration of the SD card endurance test is completed successfully. As shown in Fig. 8b, 8000 write-read-verify cycles were successfully conducted in a 10-h-long endurance test for the OBCv2. Further details of the testing and analysis are available in internal technical reports. Interfacing tests are currently being conducted between the various subsystems, which will be followed by long-duration tests and environmental tests.

5 Conclusions

The PiLOT PS4-OP payload is designed primarily for a technology demonstration of subsystems slated to fly in future small-satellite missions. The engineering models of the subsystems (OBC, TTC, RADFET board) have been realized and are currently being integrated together. The key tasks remaining are the assembly of the payload and system-level long-duration testing. Currently, simulations for the vibrational and thermal-vacuum analysis are being done to make sure that the payload will be able to survive the environmental conditions. After this, the flight model will be fabricated and will undergo environmental testing including vibrational testing and thermal vacuum testing, before its expected launch in 2021. Successful demonstration of the PiLOT mission will aid the in-house-developed OBC, TTC, and radiation monitoring RADFET payload to reach TRL-9 status so that these can be used as reliable subsystems for future small-satellite missions.

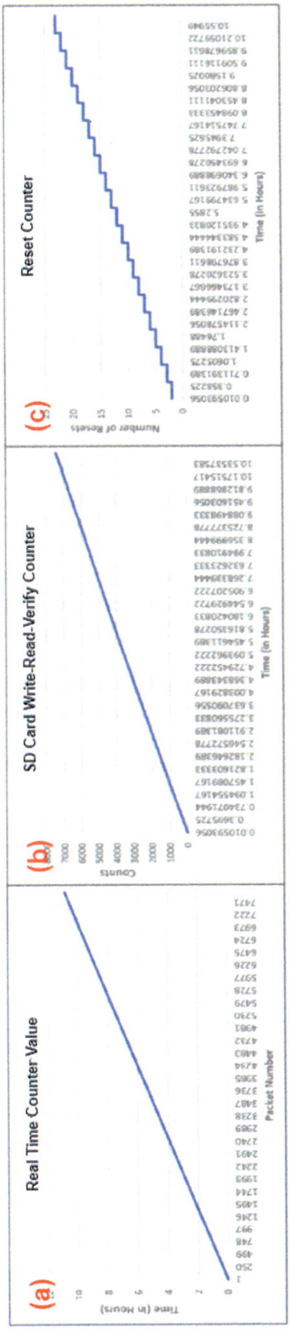

Fig. 8 OBC Testing Plots: **a** real-time counter value, **b** SD card write-read back counts, **c** number of resets

Acknowledgements We thank all the students, faculty, and staff members of the Small-spacecraft Systems and PAyload CEntre (SSPACE) for their support in the development of the PiLOT mission. In particular, we acknowledge and thank Sanidhya Vijaywat (alumnus of IIST) for his contributions to the development of the OBC as well as the PiLOT mission.

References

1. Announcement of opportunity (AO) for orbital platform, an avenue for in-orbit scientific experiments. Indian Space Research Organisation, 2019, [ONLINE], Available: https://www.isro.gov.in/sites/default/files/orbital_platform-_ao.pdf
2. Ranade A et al (2020) Survey and analysis of payloads for missions on PSLV orbital platform, [ONLINE], Available: https://www.researchgate.net/publication/342013689SurveyandAnalysisofPayloadsforMissionsonPSLV'sOrbitalPlatform
3. PSLV-C45/EMISAT MISSION Launch Kit. Indian Space Research Organisation, Bangalore, 2019, [Online], Available: https://www.isro.gov.in/sites/default/files/pslv-c45launchkitv5.0fn23.03.19.pdf
4. Alvarez M, Manzano P, Escribano D, Hernando C, Jimenez JJ, Sampedro S, Arruego I (2016) On-orbit measurements of TID and dose rate from two RADFETs on board nanosat-1b satellite. In: IEEE radiation effects data workshop
5. TECHNICAL DATA VT02 400nm RADFET, Varadis, 2020, [Online], Available: https://www.varadis.com/wp-content/uploads/2020/06/VT02-Datasheetrev1p4.pdf
6. IGLOO2 and SmartFusion2 65nm Commercial Flash FPGAs, Interim Summary of Radiation Test Results, Microsemi, 2014
7. Boyajian S et al (2019) Inspiresat-1: an ionosphere and solar x-ray observing microsat. In: Proceedings of the 33rd annual AIAA/USU small satellite conference, session V: next on the pad, SSC19-V-06, Utah, USA
8. Duann Y et al (2020) Ideassat: a 3U cubesat mission for ionospheric science. Adv Space Res 66:01

Design and Performance Validation of CETSAT Sensor Module Fabricated Using COTS for Low Earth Orbit Application

M. Achath Vaishnav, S. Lakshmi, Pratyush Prakash, M. Gopal, Tony James, George Alappat, S. Adharsh, C. V. Jiji, Abhilash Suryan, R. R. Ajith, and Ranjith S. Kumar

Abstract In this work, design, development and testing of a sensor module for small satellite application using commercial-off-the-shelf (COTS) components was performed in connection with the student satellite programme of College of Engineering Trivandrum (CETSAT). Various initiatives to democratize space missions have been implemented all over the world. Indeed, COTS components are finding their way into small satellites lately as they lead to significant cost reductions and are easily available. It involves using industrial and automotive-grade electronics and other elements from nontraditional space markets. Such components are particularly useful in missions on a small-time frame or budget. Testing the space-worthiness and long-time durable data collection while in orbit of these COTS is an important preliminary step to be carried out prior to the launch of a full-fledged satellite. The payload constructed by College Of Engineering Trivandrum is a sensor module composed of a 3-axis magnetometer with associated breakout board as well as four temperature sensors to obtain the attitude and temperature information, respectively. The deployment of the magnetometer also helps in confirming the effectiveness or lack thereof in the magnetometer-only approach. A ground station was designed and set up to

M. A. Vaishnav (✉) · S. Adharsh · C. V. Jiji
Department of Electronics and Communication, College of Engineering Trivandrum, Thiruvananthapuram, Kerala 695016, India
e-mail: vaishnavma@cet.ac.in

S. Lakshmi · P. Prakash · T. James · A. Suryan · R. R. Ajith
Department of Mechanical Engineering, College of Engineering Trivandrum, Thiruvananthapuram, Kerala 695016, India

G. Alappat · R. S. Kumar
Department of Electrical and Electronics Engineering, College of Engineering Trivandrum, Thiruvananthapuram, Kerala 695016, India

M. Gopal
Department of Computer Science and Engineering, College of Engineering Trivandrum, Thiruvananthapuram, Kerala 695016, India

R. S. Kumar
CET Centre for Interdisciplinary Research, College of Engineering Trivandrum, Thiruvananthapuram, Kerala 695016, India

© The Author(s), under exclusive license to Springer Nature Singapore Pte Ltd. 2023
R.S. Priyadarshini and T. Sundararajan (eds.), *Advances in Small Satellite Technologies*, Lecture Notes in Mechanical Engineering, https://doi.org/10.1007/978-981-19-7474-8_10

receive data from the satellite module and calibrate its performance with the data obtained by the ground station of Indian Space Research Organization. The chosen sensors were passive sensors and it is hypothesized that such sensors have a lower chance of failure in the Thermovac and Vibration tests as well as in the Low Earth Orbit (LEO). The details of the design, fabrication and testing are presented in this paper and inferences of the performance of COTS sensors in simulated environments of LEO were made.

Keywords Small satellite · COTS · LEO · Thermovac · Vibration testing

1 Introduction

CETSAT is the Student Satellite Project of College Of Engineering, Trivandrum. The government of Kerala, in 2016 sanctioned a new student satellite project for College of Engineering Trivandrum. This project aims to design, develop, fabricate and launch a nano satellite in association with ISRO and CET Centre for Interdisciplinary Research (CCIR) is the nodal agency. CETSAT aims to achieve a series of student satellites, with each being more sophisticated and employing newer technologies than the predecessors. Over 10 student satellites have been successfully launched from India under the guidance of ISRO [1–3].

The use of commercial grade components in satellites especially gained traction in the late 2000s [4]. Swisscube, a small satellite developed by EPFL (Ecole Polytechnique Fédérale de Lausanne) space centre, was one of the earliest satellites made from commercial grade components which completed the mission parameters. Swisscube was launched in 2009 and had an expected life cycle of 3 months, and is still functional after 10 years [5]. COTS usage in satellites gained public attention when NanoSatisfi LLC, founded by three graduates from International Space University started a kickstarter in 2012 to put a nano satellite, ArduSAT in space by 2013. ArduSAT was a nano satellite of size 1U, made from hobby kit components [6].

Raspberry Pi, another popular Single Board Computer (SBC), has been used in many nano satellites [7] and in some small satellites as the main control board such as the Autonomous Assembly of a Reconfigurable Space Telescope (AAReST)—a major satellite venture started in 2009 by California Institute of Technology (Caltech), the University of Surrey—Surrey Space Centre (SSC) and the Indian Institute of Space Science and Technology (IIST). The satellite is controlled by several Raspberry Pi Computer modules [8] and aims to build a large space telescope by a swarm of small mirror satellites. With the advancements in manufacturing processes, most commercial electronic components have small failure rates. A commercial grade component when shielded properly in a well-constructed satellite frame is less likely to fail as demonstrated by the success of ArduSAT and Swisscube missions.

In the similar lines, many ventures and startups have started launching small satellites in India. SpaceShare is a joint initiative of ISRO and Exseed Space to provide free experimental space missions in order to encourage participation in India's Space Programme. It aims at democratization of space by removing many technology, cost and time hurdles by hosting a shared platform for power and communications among various payloads.

The SpaceShare programme is envisaged to launch payloads of ten organizations in a common bus at the Low Earth Orbit (LEO) with Polar Satellite Launch Vehicle Stage 4 (PSLV)-Orbital Platform (PS4-OP). The payload was among the ten payloads launched by various colleges, universities and NGOs in the SpaceShare mission. Utilizing this opportunity, students of College of Engineering Trivandrum (CET) designed and fabricated a module in house composed of COTS components to test their performance in LEO and to obtain valuable temperature and attitude information which will be useful for the design and development of the student satellite initiative of College of Engineering Trivandrum. In this paper, the design, analysis, fabrication and testing of electronic and mechanical components were discussed.

2 Payload System Description

The sensor module system details are detailed below.

2.1 Payload Objective

The broad objective of the CETSAT-SpaceShare payload (CSP) was to launch a sensor module with capabilities including temperature sensing, attitude information sensing and generation of a pre-programmed pulse signal.

The specific objectives are enlisted below.

(1) To design and develop a sensor module consisting of COTS components.
(2) To gather scientific understanding in the various testing processes involved with space qualification of a satellite component and analyse data.
(3) To obtain temperature and attitude information in orbit and to evaluate the quality of data received at the ground station.

2.2 Orbit Specifications

The orbit specified for the mission was the Low Earth Circular Orbit at an altitude of 400–500 km from sea level which is the same as that used by ISRO for PS4-OP.

2.3 Payload Components

The CSP was composed of the following sensor and controller components:

(1) a tri-axial magnetometer (TAM) with associated breakout board,
(2) temperature sensors (×4),
(3) microcontroller with associated circuitry, as shown in Figs. 1 and 2.

Four temperature sensors were installed to obtain data on the temperature fluctuations in a LEO satellite so as to facilitate effective thermal designs from the ground simulations. The TAM was installed to measure the attitude of the satellite regardless of its orientation in space and to infer whether the error associated with the magnetometer-alone approach is within permissible limits. It was expected to change its direction rapidly enough to make the computation of its time derivative possible and these changes during the orbit are supposedly large enough to enable the determination of the three Euler angles. Both the sensors incorporated in the payload are passive in nature, i.e. the sensing element of both the temperature sensor and magnetometer are resistive. Generally, passive sensors are less prone to failure than active sensors. The communication between the components of the payload were according to the I2C (inter-integrated circuit) protocol and that between the payload

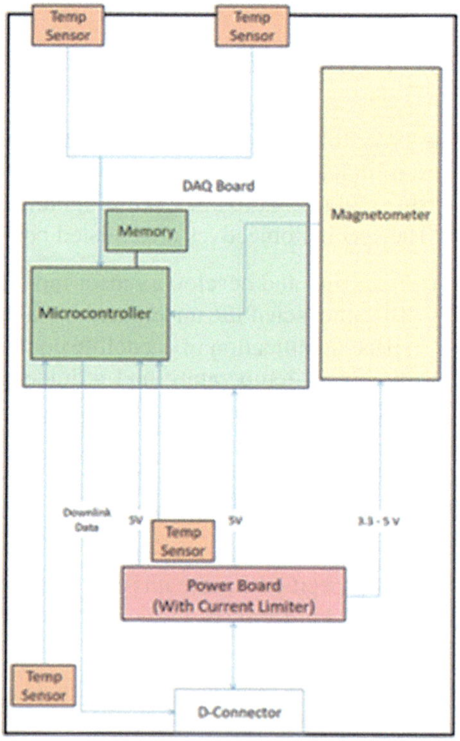

Fig. 1 Illustration of payload components and their connections

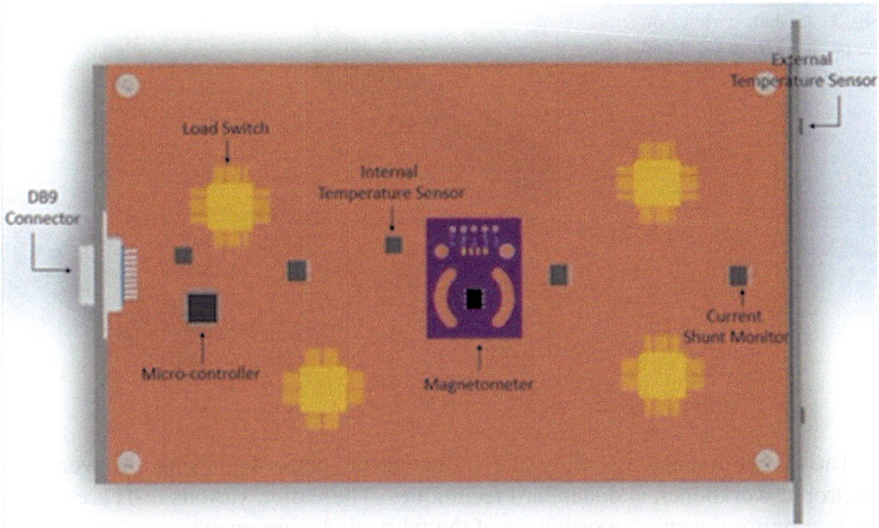

Fig. 2 Top view of the payload

and the SpaceShare interface was according to the UART (Universal Asynchronous Receiver-Transmitter) protocol.

Magnetometer. Magnetometer MAG3110 from Honeywell was chosen off-the-shelf for the mission as it satisfied the major three requirements, digital measurement in three axes, availability of exclusive circuit board and power consumption within permissible limits (< 1 W). An additional breakout board from Sparkfun electronics was included with MAG110. It is a small printed circuit board (PCB) equipped with various electrical components to make the integration of various chip types easier. The module includes a state-of-the-art, high-resolution HMC118X series magneto-resistive sensor, plus an ASIC containing amplification, automatic degaussing strap drivers, offset cancellation and a 12-bit ADC (analog to digital converter) that enables 1–2 compass heading accuracy. The I2C serial bus allows for an easy interface. The 12-Bit ADC coupled with low noise AMR sensors achieves 5 milli-gauss field resolution in 8 Gs fields.

Temperature sensor. A silicon-based thermistor with a positive temperature coefficient (PTC), TMP102 was chosen to sense temperature within the satellite in the LEO. It has a wide operating temperature ranging from 55 to +125 °C which is the desired operating range. It offers robust performance due to the immunity offered to environmental variation and built-in fail-safe behaviours at high temperatures.

Microcontroller. The ATMEGA128 microcontroller which is a low-power CMOS 8-bit microcontroller based on the AVR enhanced RISC (Reduced Instruction Set Computing) architecture with 128 kbytes In-System Programmable Flash was mounted on the PCB. This was done to study the performance of the ground station constructed at CET. The assembled view of the components are available in the CAD drawing as in Fig. 3.

Fig. 3 Isometric views of the payload from front and rear side

The fabrication and assembly of components are performed using a conventional electronic workbench and standard techniques. Thereafter, various tests are carried out to monitor the space worthiness of the fabricated components.

3 PSLV Stage 4-Orbital Platform (PS4-OP)

The PSLV Stage 4-Orbital Platform (PS4-OP) is a novel idea formulated by ISRO to use the spent 4th stage of the PSLV to carry out in-orbit scientific experiments for an extended duration of 1–6 months. Our CSP payload will be housed in the PS4-OP within the SpaceShare Chassis. SpaceShare is a joint initiative of ISRO and Satellize (Exseed Space Innovations Pvt. Ltd.) to provide free experimental space missions in order to encourage the participation of non-commercial entities such as Indian universities, colleges, research organizations and NGOs in India's Space Programme. The SpaceShare programme employs a SpaceShare Chassis that houses up to 10 payloads. The CETSAT payload is one of the payloads within the chassis. Each payload is on a 16 cm × 10 cm Eurocard-sized circuit board. It is provided with 5 V, 1A (5 W power supply) and a 9.6 Kbps serial data link to the ground.

Each payload 'card' will be inserted into one of the ten slots, and fastened by using screws. The entire chassis is mounted to the PS4-OP. The isometric views of the SpaceShare Chassis is shown in Fig. 4.

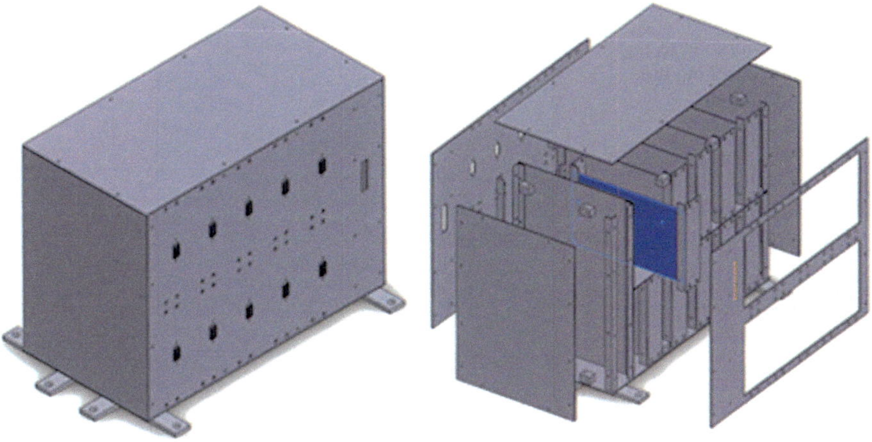

Fig. 4 Isometric views of the SpaceShare chassis (Courtesy—Satellize, Exseed Space Innovations Pvt. Ltd.—satellize.com)

4 Test Details

4.1 Vibration Test

The vibration test of the payload was performed at the Vibration and Shock Test section of ISRO Inertial Systems Unit as per the Satellize SpaceShare guidelines.

The following vibration tests were carried out in three axes:

(1) Low level Sine 0.5 G pre and post resonance survey
(2) Sine Burst 17.6 G tests at 1/3 of first fundamental mode frequency of PCB, dwelled for 3 s (10–12 number per cycle)
(3) PSLV qualification-level random vibration test at 13.5 gRMS for 120 s

During the test, vibration performance was measured on PCB near critical components like the microcontroller, temperature sensors and crystal oscillator online as given in Fig. 5.

4.2 Thermovac Test

Next, the Thermovac test was performed with the PCB in passive mode for the entire duration of the test and no abnormal observations were made. The vacuum gauge is of EDWARDS make and is working in full range. The details of the data acquisition system is available in Table 1. The online monitoring of thermal data is recorded for all the four sensors. Next, the thermocycling is performed to test the thermal stability

Fig. 5 Image of CSP mounted on vibration testing device while checking the mechanical integrity

Table 1 Temperature data acquisition system details

Sl. No.	Temperature sensor used	Type	ID No. in chamber	Name of channel	Type of channel	Sensor ID No. in data acquisition system
1	Thermocouple	T	TC-4	Base plate-1 (BP-1)	Control	CH-104
			TC-5	Base plate-2 (BP-2)	Control	CH-105
			TC-2	Annular shroud-1	Monitoring	CH-102
			TC-3	Annular shroud-2	Monitoring	CH-104

of the assembled component as shown in Fig. 6. Short cold and short hot soaking were performed for 2 h at 1 × 10–5 mbar pressure settings, see Table 2.

5 Discussions

5.1 Vibration Tests

(1) The results of 0.5 g Sine Resonance survey is shown in Fig. 7. The first mode frequency of the PCB card from the 0.5 g Sine Resonance survey was observed to be at about 249 Hz along the Thrust axis (+Z).
(2) The sine resonance survey conducted along the other two axes ('perpendicular to connector' and 'along connector') were comparatively negligible.
(3) The vibration response plot for the PSLV DQT-level random vibration of 13.5 gRMS is shown in Fig. 8.

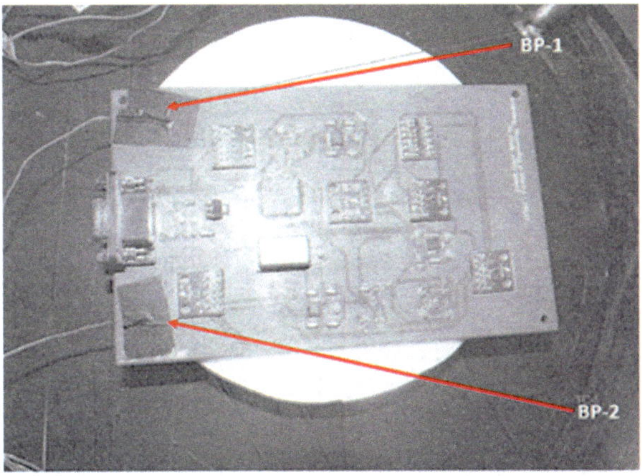

Fig. 6 Thermocouple locations on module inside thermo-vacuum chamber

Table 2 Thermo-vacuum cycling test specifications

Test	Temperature (°C)	Vacuum (mbar)	Duration (h)	No. of cycles
Short cold soak	−10	1×10^{-5}	2	1
Short hot soak	+50	1×10^{-5}	2	

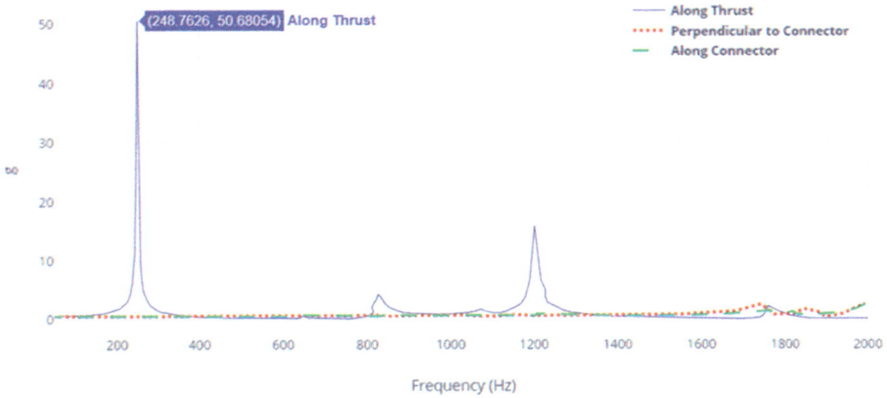

Fig. 7 Results of sine 0.5 g resonance test along thrust (ZZ) axis

(4) The highest amplitude was along the Thrust axis, observed at about 244 Hz.
(5) The results of the response to PSLV DQT-level random vibration of 13.5 gRMS along the other axes were comparatively negligible.
(6) The performance of the PCB when powered on after vibration along each axis was shown to be satisfactory.
(7) Thereby, the space worthiness of the sensor module fabricated out of COTS components are demonstrated successfully.

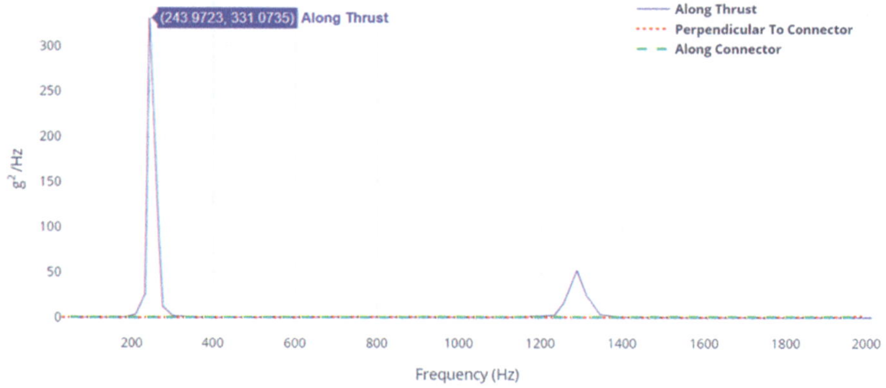

Fig. 8 Results of random 13.5 g_{RMS} vibration test

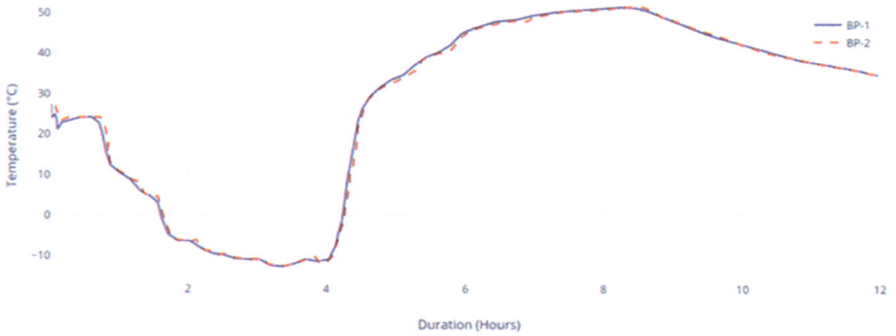

Fig. 9 Results of thermovac test

5.2 *Thermovac Tests*

The payload successfully passed the Thermovac test. Data from the two thermocouples (BP-1 and BP-2) are shown below. The location of the thermocouples is shown in Fig. 9.

6 Summary

This research work focuses on the design, fabrication and testing of a sensor module for CETSAT student small satellite project. The payload consists of a magnetometer and thermistors and is designed to be included in the SpaceShare programme of Exseed Space and is expected to perform experiments on PS4-OP. The mechanical design was performed initially considering the mechanical structure of electronics components. The module is fabricated using standard assembling procedure using COTS components. From the results of the Thermovac and Vibration tests on the payload deployed to test the performance of COTS components in LEO missions, it can be inferred that COTS components are indeed a viable option for space missions in the LEO as they passed the ground tests meant for such a mission. Differing from our hypothesis that passive sensors are less likely to fail in comparison to active sensors, the payload passed the test only when the initial oscillator crystal was replaced by an active oscillator crystal which included an active oscillator circuit associated with the crystal.

Acknowledgements This research was supported by the Government of Kerala, India and the authors acknowledge the same. The authors sincerely appreciate the technical support by the ISRO-Inertial System Unit, Trivandrum, Kerala in performing the vibration and Thermovac tests. The authors thank Exseed Space Pvt. Ltd. for accommodating the payload in their satellite missions.

References

1. Angadi C, Manjiyani Z, Dixit C, Vigneswaran K, Avinash GS, Narendra PR, Prasad S, Ramavaram H, Mamatha RM, Karthik G, Arpan HV, Sharath AH, Sashi Kiran P, Visweswaran K (2011) STUDSAT: India's first student Pico-satellite project. In: 2011 aerospace conference, Big Sky, MT, USA, pp 1–15. Homepage, https://doi.org/10.1109/AERO.2011.5747469
2. Prashanth G, Kumar A, Mulay S (2013) Pratham satellite: faraday rotation-based TEC measurement. Indian J Radio Space Phys (IJRSP) 42(3):197–203
3. Murugan P (2020) Satellite projects by Indian students. Int J Eng Res Technol (IJERT) 9(3):551–558
4. Burleigh SC, De Cola T, Morosi S, Jayousi S, Cianca E, Fuchs C (2019) From connectivity to advanced internet services: a comprehensive review of small satellites communications and networks. In: Wireless communications and mobile computing. Wiley Hindawi Publications, pp 1–18. Homepage, https://doi.org/10.1155/2019/6243505
5. Borgeaud M, Scheidegger S, Noca M, Roethlisberger G, Jordan F, Choueiri T, Steiner N (2010) SwissCube: the first entirely-built swiss student satellite with an earth observation payload. In: Small satellite missions for earth observation. Springer, Berlin, Heidelberg, pp 207–213. Home page, https://doi.org/10.1007/978-3-642-03501-2_19
6. Geeroms D, Bertho S, De Roeve M, Lempens R, Ordies M, Prooth J (2015) ARDUSAT, an Arduino-based CubeSat providing students with the opportunity to create their own satellite experiment and collect real-world space data. In: Proceedings ARDUSAT, pp 1–5. Belgium
7. Lal B, Blanco E, Behrens J, Corbin B, Green E, Picard A, Balakrishnan A (2017) Global trends in small satellites. Technical Report, IDA Science and Technology Policy Institute, p 8638
8. Underwood C, Pellegrino S, Lappas VJ, Bridges CP, Baker J (2015) Using CubeSat/micro-satellite technology to demonstrate the autonomous assembly of a reconfigurable space telescope (AAReST). Acta Astronautica 114:112–122. Homepage, https://doi.org/10.1016/j.actaastro.2015.04.008

Systems for Small Satellites

Design and Development of Cold Gas Propulsion System for Smart Space Robot

P. Arunkumar, G. Mahesh, B. Ajith, Ebin Thomas, and Aishwarya Shankhdhar

Abstract Future human-space exploration efforts aim to achieve maximal synergy between human and robotic missions, where in human-scale robots shall supplement astronaut exploration activities and also undertake robotic precursor missions. Towards this, design and development of a flying Smart Space Robot (SSR) is initiated. The Smart Space Robot (SSR) is a space flying robot which will be tethered to the unmanned orbital platform on the fourth stage (PS4-UOP) of Polar Satellite Launch Vehicle (PSLV) for micro g experiments. SSR is of nano-satellite class with dimensions of ~350 × 350 × 350 mm, weight of ~12 kg and power of ~30 W. The nanosat has to undergo different phases of operations which include deployment, station keeping, retrieval and docking. These manoeuvres calls for a robust system capable of operating for an extended duration with multiple restart and pulsing capability. This paper details out the development and design process of a cold gas system within the constraints of space, volume, mass, voltage and power as designated by the nanosat specification which meets the mission requirements.

Keywords Nanosat. PSLV · Orbit manoeuvre

1 Introduction

While considering small satellites for a wide range of earth orbit and even interplanetary missions, cold gas propulsion systems are quintessential. Cold gas systems are excellent where a low total impulse is required with not much bothering about specific impulse. These systems are often used in small satellites since 1960's [1]. It has proven to be the most suitable and successful low thrust space propulsion for Low

P. Arunkumar (✉) · G. Mahesh · B. Ajith · E. Thomas · A. Shankhdhar
Pressure Fed Engines and Thrusters Group, Liquid Propulsion Systems Centre, Valiamala, Trivandrum 695547, India
e-mail: arun_uthradom@yahoo.co.in

E. Thomas
e-mail: tomas.ebin4@gmail.com

© The Author(s), under exclusive license to Springer Nature Singapore Pte Ltd. 2023

R.S. Priyadarshini and T. Sundararajan (eds.), *Advances in Small Satellite Technologies*, Lecture Notes in Mechanical Engineering, https://doi.org/10.1007/978-981-19-7474-8_11

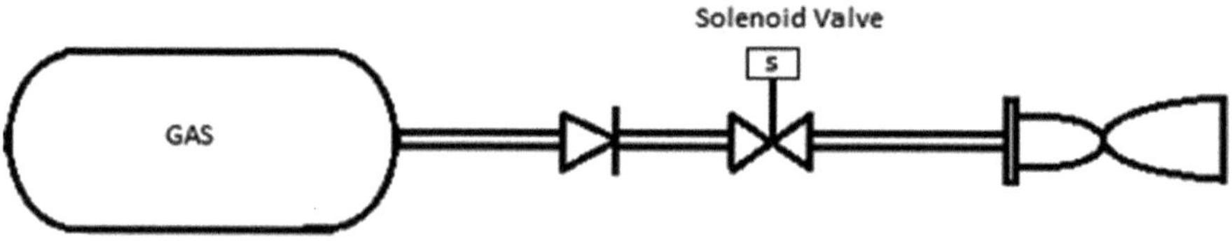

Fig. 1 Schematic of cold flow propulsion system

Earth Orbit (LEO) manoeuvres, due to its low complexity, efficient use of propellant which presents no contamination and thermal emission besides its low cost and power consumption. The notable advantages obtained from cold gas systems are low budget, mass and volume. The system mainly consists of a propellant tank, solenoid valves, thrusters, tubing and fittings [2] (Fig. 1).

The propellant tank is used to store the fuel required for attitude and orbit control of satellite during its operation. The propellant used in cold gas systems is compressed gas. Thrusters provide sufficient amount of force to provide stabilisation in pitch, yaw and roll movement in case of satellite and can also impart to or kill the velocity of spacecraft in translational direction. From the perspective of design, three components that play an important role of cold gas propulsion systems are mission design, propellant storage and cold gas thruster [2].

2 Design and Development

The development of a thruster is multifaceted optimisation problem taking into consideration the manoeuvers required in the mission. The system has to perform within the constraints of limited space, weight and power and convert the energy stored as pressure into kinetic energy of the nanosat. For the same, the design stage is initiated by taking the requirements of the mission and designing a system within the constraints and bounds of the mission.

2.1 Selection of Working Fluid

It is of utmost importance to select which gas to use as the propellant. This affects component selection in terms of sizing and materials selection. Various gases have been used for cold gas propulsion like helium, carbon dioxide, butane, air and nitrogen. Selection between the several possible gas options was based on many different factors. Chief among them were inflammable/non-reactivity, availability of the gas and material compatibility [3]. The options were narrowed down to four gases that were considered in greater detail, as outlined below.

Carbon-dioxide (CO_2)

In systems where carbon-dioxide is used, it is stored in mixed liquid and gas phase. As the gaseous CO_2 is used as propellant, the liquid CO_2 evaporates to replenish it, keeping the pressure inside the propellant tank constant until all the liquid evaporates [4]. This helps in a constant tank pressure and stable thruster performance, and storage in liquid phase means that a relatively large mass of propellant can be carried with simple and light systems' design.

Helium (He)

Helium is one of the best cold gas propellant on account of its high-specific impulse. It has an Isp of approximately 180 s, which is very high for an inert gas propellant [5]. However, due to its low density, helium should be stored at very high pressure or requires a storage tank of higher volume. This often results in a heavier propellant tank and durable, leak proof plumb lines capable of supporting high pressures. Thus, the heavier system counteracts the gains from the high Isp. Furthermore, helium systems are especially prone to leaking on account of its low molecular mass. This requires specific materials driving up the cost of propulsion system.

Nitrogen (N_2)

Nitrogen gas has only a mediocre Isp of 60–80 s. It is strongly outclassed by helium in terms of efficiency. However, nitrogen is denser and less susceptible to leaking in comparison to helium which means the system design is lighter and less complex, and it is also cheaper. It is a common choice for cold gas propulsion systems in around the world.

Butane (N2)

Butane thrusters have similar ISP to GN2 thrusters but require lower storage envelope due to its higher density and conveniently, it can be stored at a very low pressure, hence no pressure-regulation system is required. Despite lower system weight, butane is more prone to liquefaction and there is a possibility of two phase flow which will result in lower ISP. To prevent this, heaters and plenum chambers have to be provided which complicates system. The system is also prone to sloshing.

Out of the numerous options, GN2 has been selected owing to its simplicity in handling and operations and its leak proof property.

2.2 System Specification

The cold gas propulsion system is designed to impart velocity to the nanosat for various manoeuvres. The thrusters will be used in the following phases of SSR operation:

- Imparting separation velocity—along deployment direction (4 s burn)
- De-tumbling and attitude disturbance rejection

- Killing of delta V
- Azimuth correction
- Station keeping
- Retrieval.

The nanosat performs 4 major operations and the total impulse requirement was estimated to be 100 Ns. The thrust generated by the proposed thrusters are 150 mN. This results in a total firing duration of 667 s. From the operating duration, the volume of working fluid required and tank volumes can be worked out.

2.3 Design of Thrust Chamber and Propulsion System

Thrusters are the convergent-divergent nozzles (Fig. 2) that provide desired amount of thrust to perform manoeuvres in space. The nozzle is shaped such that high-pressure low-velocity gas enters the nozzle and is compressed as it approaches smallest diameter section, where the gas velocity increases to exactly the speed of sound.

Nozzle Design (First Cut Calculation)

The thruster is operated in the vacuum and the following parameters are taken as design inputs.

Design parameters—thrust-Th (150 mN); chamber pressure-P_c (7 bar); and exit pressure $-P_e$.

By assuming an exit pressure, the expansion ratio (area ratio) can be found out by using,

$$\varepsilon = \frac{A_e}{A_t} = \frac{\left(\frac{2}{\gamma+1}\right)^{\frac{1}{\gamma-1}} * \left(\frac{\gamma-1}{\gamma+1}\right)^{\frac{1}{2}}}{\left(\frac{P_e}{P_c}\right)^{\frac{1}{\gamma}} * \left[1 - \left(\frac{P_e}{P_c}\right)^{\frac{\gamma-1}{\gamma}}\right]^{\frac{1}{2}}} \tag{1}$$

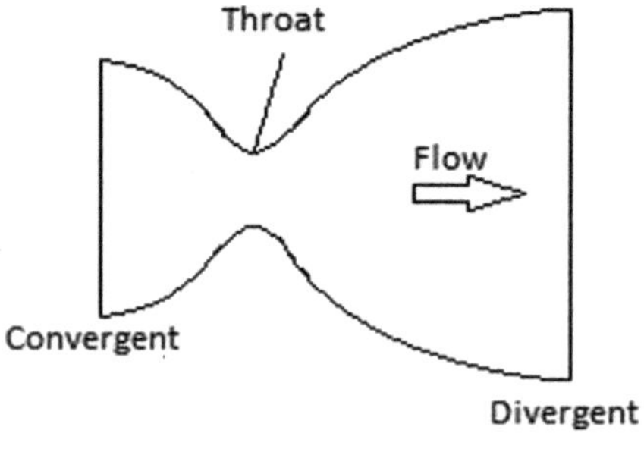

Fig. 2 Typical convergent divergent nozzle

The area ratio (ε) is limited by mechanical and envelope constraints. The ideal combination of the exit pressure (P_e) and ε is finalised after extensive iteration.

The exit Mach no (M_e) can be found by solving isoentropic relation,

$$\frac{P_e}{P_c} = \left(1 + \frac{\gamma - 1}{2} M_e^2\right)^{\frac{-\gamma}{\gamma-1}} \quad (2)$$

The thrust coefficient (C_f) can be found by solving the equation,

$$C_f = \sqrt{\frac{2\gamma^2}{\gamma - 1}\left(\frac{2}{\gamma + 1}\right)^{\frac{\gamma+1}{\gamma-1}}\left[1 - \left(\frac{P_e}{P_c}\right)^{\frac{\gamma-1}{\gamma}}\right]} + \left(\frac{P_e - P_{\text{atm}}}{P_c}\right) * \varepsilon \quad (3)$$

Applying the above results in the thrust equation will yield the throat area (A_t)

$$T_h = P_c * A_t * C_f \quad (4)$$

The mass flow ($\dot{m}$) required for each thruster can be found by solving the chocked flow equation

$$\frac{\dot{m}}{A_t} = \frac{\Gamma * P_c}{\eta C d \sqrt{\gamma R T_c}}$$

where, r is the Van-Kirchhoff function

$$\Gamma = \lambda * \left(\frac{2}{\gamma + 1}\right)^{\frac{\gamma+1}{2(\gamma-1)}} \quad (5)$$

All other major design parameters of the nozzle like throat dia (D_t) and exit dia (D_e) can be found by solving Eq. (1).

From the mass flow and the total duration of firing, major system parameters like the total volume of working fluid, tanks capacity, and plumb lines can be finalised.

Nozzle Design (CFD Route)

The nozzle design is also deduced using CFD tool. The results are satisfactory with the thrust imparted by the nozzle which is on the higher side with 180 mN as compared to 150 mN.

- Geometric Modelling and Meshing
- Actual 2-D model of Cold Gas thruster and far-field (15 d horizontal and 5 d vertical) created in AUTOCAD
- AUTOCAD model imported to POINTWISE
- Approx. 35,000 nos. nodes considered for solving the problem
- Mesh file was imported to FLUENT for CFD analysis (Fig. 3).

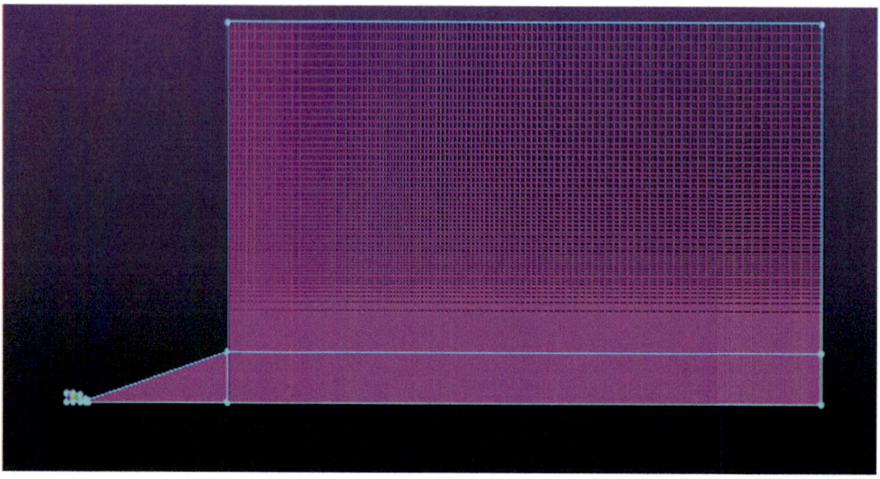

Fig. 3 Cold flow thruster meshed in POINTWISE

Governing Equations

(1) Mass Conservation Equation

$$\frac{\partial \rho}{\partial t} + \frac{\partial(\rho u)}{\partial x} + \frac{\partial(\rho v)}{\partial y} + \frac{\partial(\rho w)}{\partial zx} = 0 \tag{6}$$

(2) Momentum Conservation Equation

$$\rho\frac{Du}{Dt} = -\frac{\partial \rho}{\partial x} + \frac{\partial}{\partial x}\left[2\mu\frac{\partial u}{\partial x} + \lambda \text{div} u\right] + \frac{\partial}{\partial y}\left[\mu\left(\frac{\partial u}{\partial y} + \frac{\partial v}{\partial x}\right)\right] + \frac{\partial}{\partial z}\left[\mu\left(\frac{\partial u}{\partial z} + \frac{\partial w}{\partial x}\right)\right] + S_{Mx} \tag{7}$$

(3) Energy Conservation Equation

$$\rho\frac{DE}{Dt} = -\text{div}(\rho u) + \left[\frac{\partial(u\tau_{xx})}{\partial x} + \frac{\partial(u\tau_{yx})}{\partial y} + \frac{\partial(u\tau_{zx})}{\partial z} + \frac{\partial(v\tau_{xy})}{\partial x} + \frac{\partial(v\tau_{yy})}{\partial y} + \frac{\partial(v\tau_{zy})}{\partial z} + \frac{\partial(w\tau_{xz})}{\partial x} + \frac{\partial(w\tau_{yz})}{\partial y} + \frac{\partial(w\tau_{zz})}{\partial z}\right] + \text{div}(\text{k grad}\, T) + S_E \tag{8}$$

Method of solution

- Density based Solution
- 2-D Axisymmetric
- SST K-Ω turbulence model.

Table 1 CFD results

Sl. No.	Parameters	Results
1	Thrust (N)	0.153
2	Mass flow rate (mg/s)	224
3	Exit pressure (Pa)	21
4	Exit Mach No	10.3
5	Cd at throat	0.96

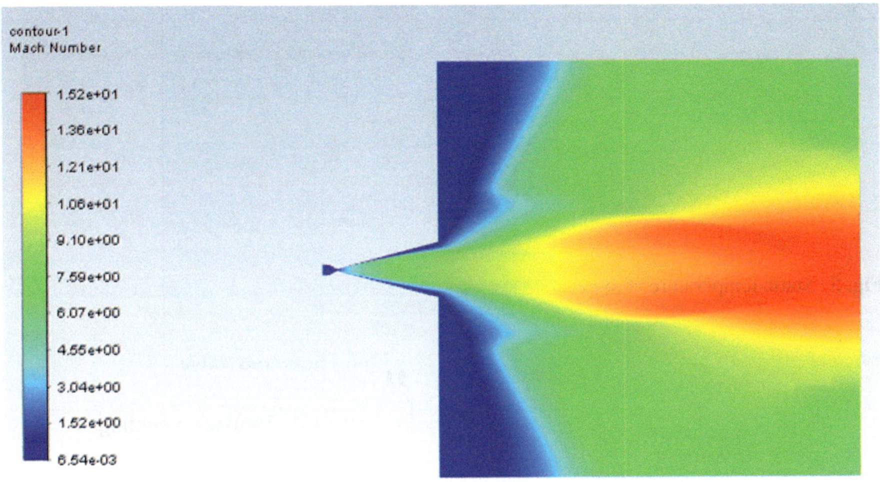

Fig. 4 Mach number contour

The CFD analysis results are shown in Table 1, and the contours are shown below (Figs. 4 and 5).

2.4 *Thruster Valve*

Thruster valve controls the operation of the thrusters. The valve must be robust and capable of multiple operations so as to fulfil various mission operations. Figure 6 shows the valve assembly together with the thrust chamber. The valve is an ON/OFF solenoid valve. The valve operates by opening and closing an armature using the magnetic flux generated by a solenoid coil.

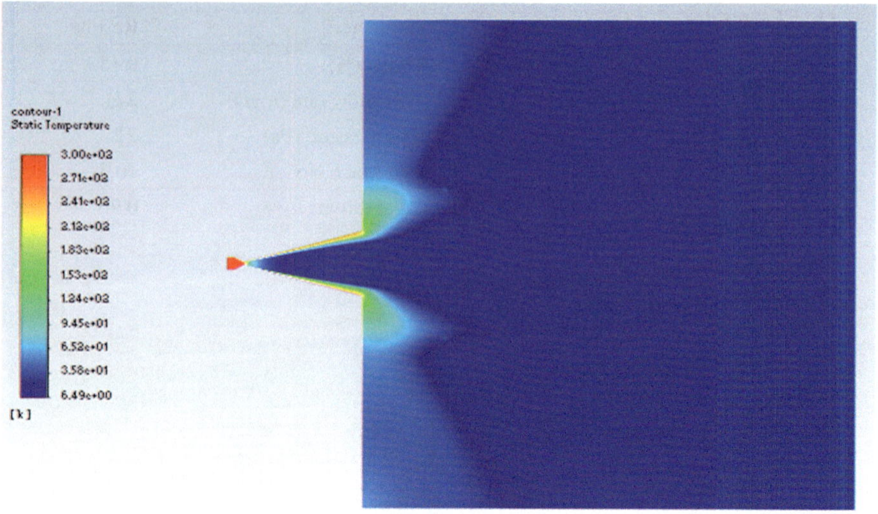

Fig. 5 Static temperature contour

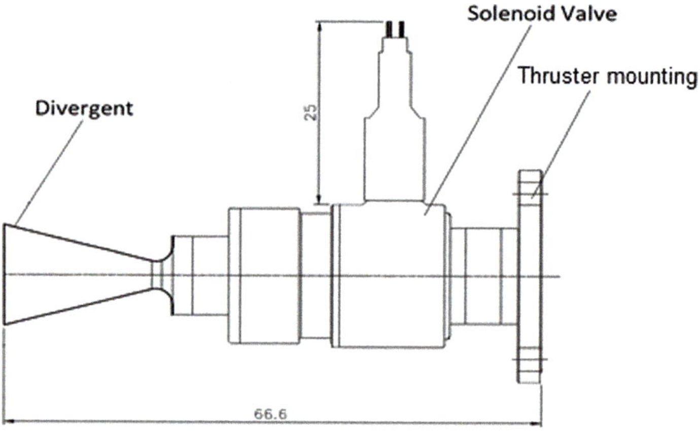

Fig. 6 Thruster assembly

2.5 *Propulsion System Design and Configuration*

The propulsion system stores the working fluid in pressurised condition and delivers it at reduced pressure to the thrusters when they are operated.

Mass of the gas required can be calculated from the total duration and mass flow per thruster.

$$\text{Total Propellant Mass } (Mt) = \text{Mission duration } (t) * \text{flow rate } (\dot{m}) \tag{9}$$

The total mass of propellant required is approximately 150 g. The tanks and plumb lines will be pressurised to a pressure of 70 bar before the mission and the thrusters will be operated at blowdown mode. The total volume required when the gas is stored at a pressure of 70 bar is as follows:

$$\text{Total Gas Volume},\ \ Vt = 1929.2\ \text{cm}^3$$

The propellants are generally loaded into the satellite with a loading factor of 1.3. The loading factor signifies the amount of propellant in the tank. A loading factor of 1.3 is imperative to an extra mass by 30% in the tank than required for the mission. This extra mass will be utilised to impart the pressure for the gas till the end of life of the satellite. Hence, the total mass of propellant required is 200 g and the volume of tank required for SSR is as follows:

$$\text{Required Gas Volume},\ \ Vt = 2440\ \text{cm}^3/2.44\ \text{ltr}$$

A singular propellant tank of 2.6 L has been conceived to cater the requirements of cold gas thrusters. The tank will be mounted centrally in the SSR.

Pressure regulation

The thrusters will be operated on regulated mode. Electronic pressure regulator (EPR) is sought after to regulate the downstream pressure irrespective of the upstream pressure. The upstream pressure varies from 70 to 35 bar during the course of the mission but the EPR will regulate the downstream pressure to 10 bar. An EPR consists of a piezo-electric valve with a pressure transducer downstream to provide the closed loop feedback system to ensure a regulated downstream pressure. In the present case of Cold Gas propulsion system, the EPR imparts desired regulated pressure of 10 bar which is critical in ensuring constant thrust of thrusters.

Latch Valve

A latch valve is also introduced downstream of EPR to arrest any possibility of leak to happen. This ensures that the FCV which is upstream of thrusters will not experience undesired pressure of 70 bar because the FCV is designed for an MEOP of 10 bar.

The finalised configuration of propulsion system is shown in Fig. 7. The system employs an electronic pressure regulator (EPR) which is to regulate the downstream pressure irrespective of the upstream pressure. A latch valve (LV) is introduced downstream of EPR to arrest any possibility of leak to happen. This ensures that the FCV which is upstream of thrusters will not experience undesired pressure of 70 bar because the FCV is designed for an operating pressure of 10 bar. The volume between the EPR and latch valve is maintained very less, in case of a pressure rise due to the leak; the large downstream volume will negate the high-pressure rise at the FCV inlet (Fig. 8).

The thrusters are configured around the thrusters in order to cater to its various manoeuvring requirement and only four thrusters can be operated at a time due to various constraints placed due to flow and power.

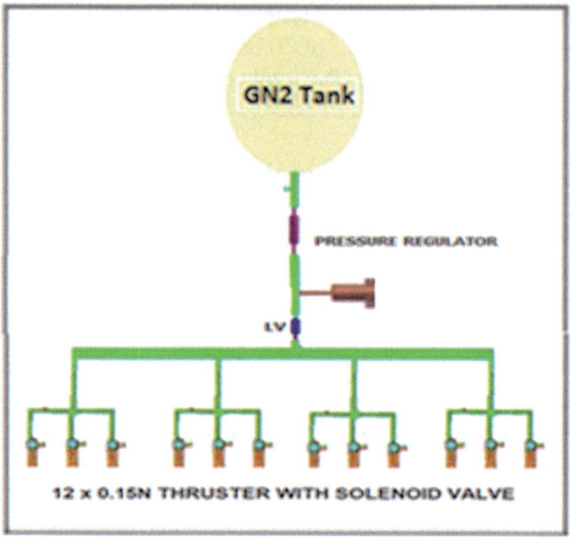

Fig. 7 Propulsion system schematic

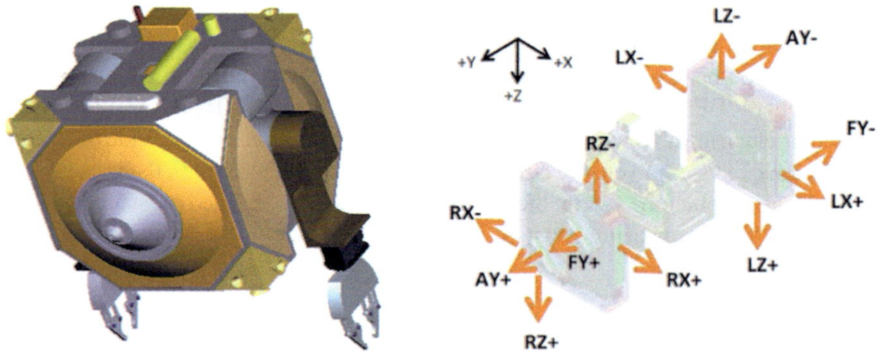

Fig. 8 Configuration of thrusters around SSR

3 Criticalities

At the onset of design phase, many criticalities were faced by the design team. The constraints placed by space, mass, low power and pressure called for devolvement of new systems and major modifications on existing system. Existing propulsion components used were carefully analysed and trade-offs were studied to meet project requirement. The major criticalities are discussed below,

3.1 Space and Mass Management

The initial available volume for propulsion system was 1 L. While considering a loading factor of 2, this called for propellants to be stored at 400 bar pressure. But after rigorous reconfiguration of components and redesign of the subsystems, the space was increased to 2.6 L. In addition, it was decided that a loading factor of 1.3 would be adequate since the mission duration is only 667 s. This reduced the pressures to 70 bar which is a comfortable value and gave confidence to the system designers.

Mass of the system is also an important constraint. Though the initial mass requirement specified a max of 3 kg for propulsion system, with the introduction of pressure regulation and pressure measurements, the mass of the propulsion system has increased to 5 kg. The trade-off between mass and reliability of the system was analysed, and it was decided to accommodate the higher mass of propulsion system.

3.2 Power Management

The lower power generated is a major constraint when designing system for nanosat. The power requirement of available solenoid valves was higher that the power supply capability. Amelioration of the power supply and miniaturisation of existing solenoid valve enabled to scarcely meet the power requirement.

3.3 Pressure Management

The constraint on the tank pressure indirectly comes from space constraint. The initial volume of 1 L was increased to 2.6 L after reconfiguration of subsystems, and storage pressure was reduced to 70 bar. It was also very significant to finalise a system for pressure drop management from 70 bar in tank to 7 bar in thrusters. In this regard, two types of system were sought after: the one was with conventional multiple orifice system and the another one was a regulated system using an Electronic Pressure Regulator (EPR).

The orifice system was considered owning to its simplicity and low mass. But on careful analysis, it was found that orifice system was unsuitable because of the repercussions of choking occurrence upstream of the thruster throat. Thus, a pressure-regulated system was selected in spite of higher mass since it provides more reliable performance.

4 Conclusion

Although cold gas thrusters are simpler compared to other propulsion systems, it still contains various constraints like volume envelope, simplicity, power constraints, voltage constraints and mass which makes it a complicated optimisation problem. In addition to that, the requirement of multiple restarts calls for a robust and versatile system. This paper lists out various constrains and thought process in the design phase of the cold gas system which uses a GN2 gas as working fluid and within the operating duration of 667 s.

References

1. Escalante-Ramirez B, Remote sensing: advanced techniques and platforms
2. Anis A, Cold gas propulsion system—an ideal choice for remote sensing small satellites. NED University of Engineering and Technology, Pakistan
3. Nothnagel SL, (2009) Development of a cold gas propulsion system for the TALARIS hopper. University of Southern California
4. Dustin B (2008) System characterization and online mass property identification of the SPHERES formation flight testbed
5. Stenmark L, Eriksson AB, The Ångström Space Technology Centre, The Ångström Laboratory, Uppsala University, Box 534, SE-751 21 Uppsala, Sweden. Cold gas micro thrusters, final edition

Compensation of Drift in Ring Laser Gyros

G. S. Anish, Nisha S. Dathan, K. Usha, and S. Paul Pandian

Abstract Ring Laser Gyroscopes (RLGs) offer highly accurate angular measurement capability for inertial class applications. They offer high stability in scale factor and bias drift. The gyro being a voluminous module with at least ten specific components or interfaces which are sensitive to temperature, thermal compensation becomes a complicated problem. Further, the direct thermal measurements are limited and sensitivity factors have to be derived from available monitoring parameters, which increases the complexity of the problem manifold. Published literature is available on many compensation algorithms; however, they limit their research mostly to using temperature reading of the optical block in order to predict bias. In this paper, we study the drift variations in the gyro as a function of temperature monitored from the optical block as well as the control signals used to sustain the optimal mode of the laser including path length control signal, high frequency oscillator signal and dithering frequency variations. We applied these data on linear and polynomial regression models and obtained up to 76% and 88% reduction in bias drift, respectively. Further, the denoising filter is studied by varying its type and the window length. Training and testing sensors in INS cluster configuration gave bias drift reduction of up to 74%. All these improvements are achieved in the navigation loop cycle timing itself enabling a smooth compensation for INS applications.

Keywords RLG—ring laser gyro · Sagnac effect

1 Introduction

Ring Laser Gyroscopes provide a very reliable and accurate means of measuring angular displacement for tactical and space applications. They were first published

G. S. Anish (✉) · N. S. Dathan (✉) · K. Usha · S. Paul Pandian
ISRO Inertial Systems Unit, Vattiyoorkavu, Thiruvananthapuram, Kerala 695013, India
e-mail: mail.anishgs@gmail.com

N. S. Dathan
e-mail: nishasdathan@gmail.com

© The Author(s), under exclusive license to Springer Nature Singapore Pte Ltd. 2023

R.S. Priyadarshini and T. Sundararajan (eds.), *Advances in Small Satellite Technologies*, Lecture Notes in Mechanical Engineering, https://doi.org/10.1007/978-981-19-7474-8_12

in 1963, and since then, several variants of the gyro have emerged [1]. With the production of RLGs, theoretical formulations have improved to better match the designs. Lamb formalization [2] is one such model which assumes the presence of an electromagnetic field within the cavity which polarizes individual atoms. These polarizations are explained according to quantum theory and creates the reaction field as per Maxwell's equation. Since, quantum theory and Maxwell's equations are used to explain in the model, it is considered as a semi-classical approach. This also imparts environmental factor dependency, particularly temperature dependency to the RLGs output.

2 Drift in RLG

Ring laser Gyroscopes offer a high accuracy system with long life compared to mechanical gyros. They also have very high bandwidth and scale factor stability compared to conventional gyros. One important performance metric for RLGs is its angle bias stability. RLGs offer extremely good bias stability of the order of 0.01 degrees/hr and better. They, however, have temperature dependency due to thermally induced stress in electro-opto-mechanical interfaces.

Many publications are available in this area analysing various aspects of compensation for different kinds of RLGs. Seon et al. [3] studies the drift compensation of RLG against temperature effects and found this to be a function of temperature and temperature gradient. Yang et al. [4] discusses RBF and BP neural network implementation based on readout signals to achieve nonlinear modelling. Li et al. [5] takes this discussion further and implements multi-point temperature gradient algorithm based on PSO SVM. Weng et al. [6] discusses the optimization of RLG bias compensation in variable temperature environment and achieved compensation for trend of bias variation with temperature. Huang et al. [7] discusses the temperature modelling of a strap-down inertial navigation system. However, these papers discuss mainly about analysing bias as a function of temperature. Further, the compensation relations seem to vary depending on the construction and excitation mechanisms of the laser gyros and cannot be generalized for all types of RLGs. Further, using multi-layer neural networks can often result in an overfit of the model and the repeatability of the model may not be ensured across multiple sensors, across multiple temperature profiles and in an INS system configuration where the close proximity of other sensors changes its ambient environment.

This paper attempts to develop a bias compensation system which predicts angle bias as a function of temperature and control parameters for RLG. The algorithm should deliver angle deviations at rates suitable for navigation loop and the model should be deployable in the sensor DSP without overshooting the tight loop frequency requirement and should make use of only limited temperature data and control signals as its inputs.

In this paper, firstly, a comparison is made between the bias drift reduction for various least square regression models between training and testing data. Then the model is trained on the sensors using path length control signal, high frequency oscillator drive signal, dither frequency monitoring signals and temperature readout signals and their results are presented. These models are trained and tested using separate thermal cycling data. Then the effect of denoising filter is studied and the results for different filter window types and duration are also shown. Further, the regression model is trained for sensor in INS cluster configuration and the results are tabulated.

3 Drift Model

The relationship between bias and temperature is found to be dependent on time derivative and gradient of temperature in [3]. In case of RF-excited RLGs, this is approximated to

$$Y_{\text{error}} = \beta_0 + B(X) + \epsilon \tag{1}$$

where β_0 is the fixed component of RLG bias, ϵ is the random error and $B(X)$ is the basis function.

$$B(X) = G \times X \tag{2}$$

where G is the generator matrix and X is the input vector of temperatures and other parameters. G and X are chosen as per the type of regression model and the input parameters used for that particular training.

The experiments for various regressions, preprocessing functions and configurations and the results are discussed in next section.

4 Compensation Scheme

A block diagram of the training scheme is provided in Fig. 1 and that of the compensation scheme is provided in Fig. 2. In order to capture thermal variations, 2 temperature sensors are placed on suitable locations in the sensor. Then thermal cycling of sensor is done, and this data is taken and passed through a preprocessing stage. This stage suppresses spurious spikes and other disturbances in the data. This stage does not suppress the random noise components in angle output. In order to get small bias variations from a sea of noise, a denoising filter is implemented. This filter is implemented by means of a rolling window over the training data in order to amplify the thermal variations and attenuate the random noise components. The results for different rolling window types, and durations are presented in Tables 1, 2 and 3. As the window length increases, the computation overload increases and we were able to achieve a nearly constant improvement within the window ranges of 0.5 s to

20 s. The 0.5-s window length gave slightly better performance due to better thermal representation.

This denoised data is taken and the regression model is trained for various monitoring parameters. The temperature showed highest dependency to bias drift. However, since the temperature variation across the optical block is not uniform, one thermal sensor is found to be inadequate for monitoring. So we used two temperature sensors for monitoring.

The processed data is then fed to the training algorithm which trains the regression model. The trained coefficients are then loaded and applied on testing data in order to get bias compensated output. In order to evaluate the sensor performance in high thermal fluctuation situations, thermal cycling is conducted. The regression model is then applied on the sensors in INS cluster configuration. The model is shown to hold for system configuration also in spite of the temperature gradient changes in INS cluster configuration. The model is further tested on short term stability data.

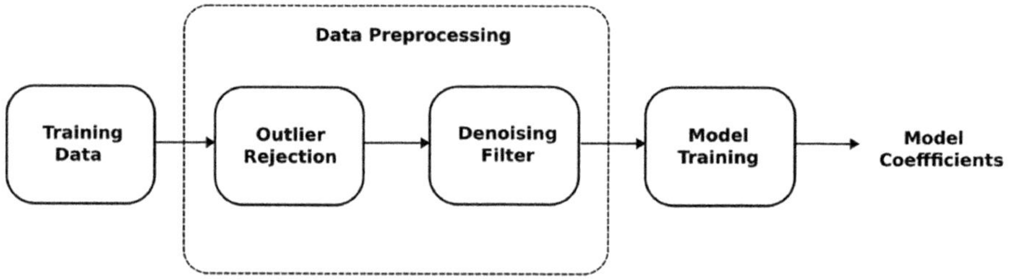

Fig. 1 Block diagram of regression training scheme

Fig. 2 Block diagram of compensation scheme

Table 1 Angle bias improvement (%) for sensor data for different LS algorithms

Sensor	L.S. linear regression (% Improvement)	L.S. polynomial regression (degree 2) (% Improvement)	L.S. polynomial regression (degree 3) (% improvement)	L.S. polynomial regression (degree 4) (% improvement)
S1	76.53	75.55	74.90	74.32
S2	67.85	82.51	84.61	88.47
S3	54.93	59.2	65.85	66.75
S4	40.11	42	55.53	48.98

Table 2 Angle bias improvement (%) for sensor data for different parameters

Sensor	Temp (% improvement)	Dither frequency (% improvement)	Temp, PLC (% improvement)	Temp, dither freq, HFO control (% improvement)
S1	76.53	71.84	76.00	76.63
S2	67.85	62.96	70.71	67.11
S3	54.93	49.87	53.97	55.34
S4	40.11	28.27	45.77	46.61

Table 3 Angle bias improvement (%) for sensor data for Boxcar window of variable width

Sensor	Boxcar 0.5 s	Boxcar 5 s	Boxcar 20 s
S1	75.82	75.78	75.70
S2	67.83	67.83	67.82
S3	55.15	55.13	55.08
S4	39.78	39.98	39.97

5 Experimental Results

Four RLGs are taken for experimental study. The observed angle and prediction value are plotted for thermal cycling data. The 1-sigma improvements in bias for temperature inputs for least squares linear and polynomial regressions of various degrees are given in Table 1. Further on training, the least squares linear regression model with control and monitoring signals namely PLC drive, HFO drive voltage and dither frequency, the percentage bias improvements are tabulated in Table 2.

While preprocessing the data, the window function is varied and the corresponding changes in performance are presented in Tables 3, 4 and 5. The 0.5-s window provided the most promising results, and Gaussian window provided slightly better results over boxcar and triangular windows. The observed bias in case of the test run and the bias value predicted by the model for least square linear regressor are shown in Figs. 3, 4, 5 and 6.

The data of the above 4 sensors in INS cluster configurations are also analysed and the corresponding bias stability improvements are compiled in Table 6.

Table 4 Angle bias improvement (%) for sensor data for triangle window of variable width

Sensor	Triangle 0.5 s	Triangle 5 s	Triangle 20 s
S1	75.82	75.77	75.7
S2	67.83	67.83	67.82
S3	55.15	55.13	55.07
S4	39.79	39.98	39.95

Table 5 Angle bias improvement (%) for sensor data for Gaussian window of variable width

Sensor	Gaussian 0.5 s	Gaussian 5 s	Gaussian 20 s
S1	76.53	76.17	76.13
S2	67.85	67.82	67.81
S3	54.93	55.11	55.05
S4	40.11	39.49	39.37

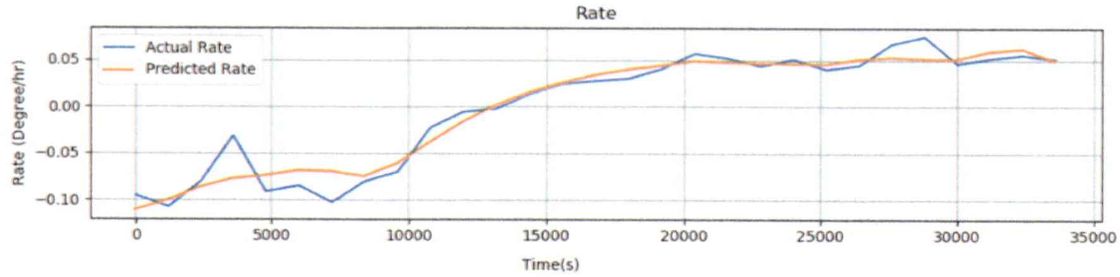

Fig. 3 Actual bias and predicted bias of sensor 1 accumulated to 1200 s

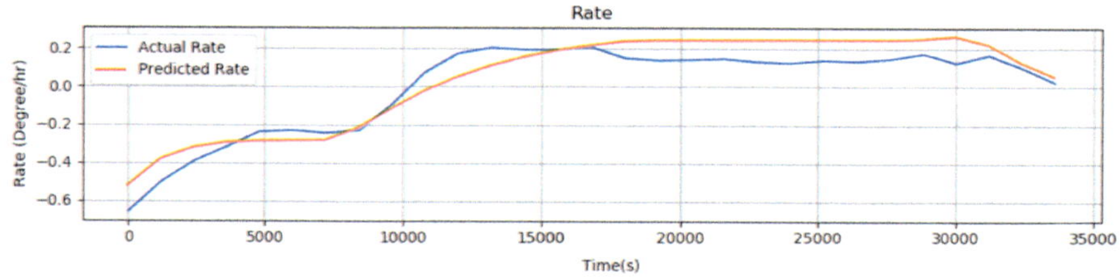

Fig. 4 Actual bias and predicted bias of sensor 2 accumulated to 1200 s

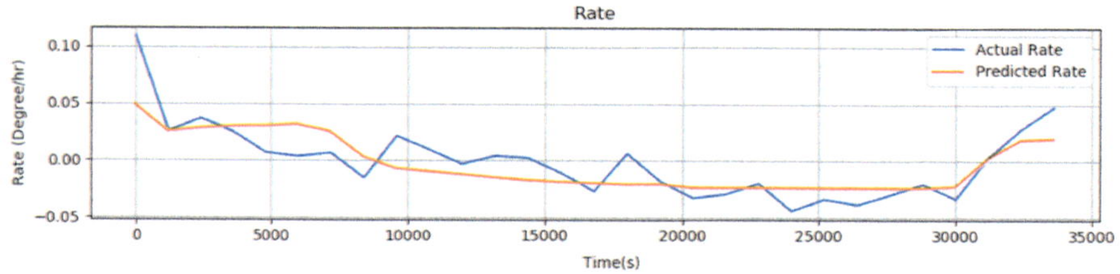

Fig. 5 Actual bias and predicted bias of sensor 3 accumulated to 1200 s

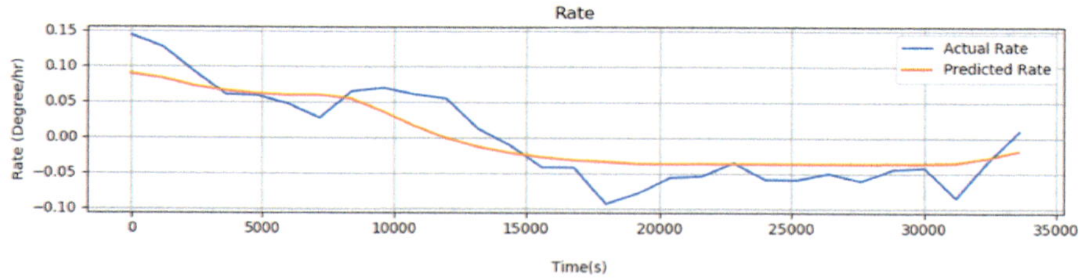

Fig. 6 Sensor 4, actual bias and predicted bias of sensor 4 accumulated to 1200 s

Table 6 Bias improvement (%) for system level short term stability data when trained with standalone sensor data and sensor data in INS cluster configuration

Sensor	Percentage improvement (training with standalone sensor data)	Percentage improvement (training with system data)
S1	48	60.1
S2	34.4	74.5
S3	60.8	47.8
S4	34.6	71.4

Further, a comparison of results obtained by training the model in sensor standalone configuration and in INS cluster configuration is also made.

6 Conclusion

The regression model offered drift improvements of up to 76% for the RLG sensor for linear regressor and up to 88% improvement for polynomial regressor. The model could track bias drift when sensor is in INS cluster configuration and offered bias improvements of up to 60% even when they are trained using standalone sensor data and showed up to 74% bias improvement in bias in INS cluster configuration when trained using the sensor data in same configuration. Based on results, bias compensation using dither frequency, HFO control voltage and path length control voltage can be used to augment the model accuracy further. The impact of different denoising filters on bias improvements and the results are also studied, and Gaussian-based filter provided better results in most cases. Multipoint temperature capturing along with more monitoring signals can be explored for future implementations.

Acknowledgements We thank Shri M Vysakh, Shri K Vijaykumar and Shri Sameer K A from ISRO for their support in understanding and analysing the data and machine learning setups.

References

1. Macek WM, Davis DTM (1963) Rotation rate sensing with travelling wave ring lasers. Appl Phys Lett 2:67–68. https://doi.org/10.1063/1.1753778
2. Menegozzi LN, Lamb WE (1973) Theory of a ring laser. Am Phys Soc 8(4):2103–2125. https://doi.org/10.1103/PhysRevA.8.2103
3. Seon H, Soo L, Soo P, Youp H, Hyun S (2008) The compensation method for thermal bias of ring laser gyro. https://doi.org/10.1109/LEOS.2008.4688823
4. Yang P, Qin Y, You J (2010) Temperature compensation for RLG based on neural network. In: Proceedings of SPIE, p 7544. https://doi.org/10.1117/12.885310
5. Li G, Zhang P, Wei G, Xie Y, Yu X, Long X (2015) Multiple-point temperature gradient algorithm for ring laser gyroscope bias compensation. Sensors https://doi.org/10.3390/s151229777

6. Weng J, Bian X, Kou K, Lian T, Optimization of ring laser gyroscope bias compensation algorithm in variable temperature environment. Sensors (Basel) https://doi.org/10.3390/s20020377
7. Huang X, Zhao L, Xu R, Yang H (2016) Research on temperature modeling of strapdown inertial navigation system. J Phys https://doi.org/10.1088/1742-6596/679/1/012014

Field-Oriented Control for Performance Improvement in Reaction Wheels and Implementation of Algorithm in FPGA

K. Ratheesh, S. Sreejith, and T. R. Haridas

Abstract The reaction wheels play a major role in attitude control of the spacecraft. As they work on the principle of conservation of angular momentum, any ripple in reaction torque will affect the performance of critical on-board instruments such as atomic clock, high-resolution cameras, etc. in the spacecraft. The three-axis stabilization system using reaction wheels aims to provide high-pointing accuracy, low jitter and long-term attitude stability. The stringent attitude and micro-vibration requirement of future missions cannot be met practically with the existing reaction wheels with conventional trapezoidal brushless direct current motor (BLDC) drive. A suitable upgrade in reaction wheels with permanent magnet synchronous motor (PMSM) instead of BLDC motor and field-oriented control instead of six-step commutation will be enabled to meet these requirements. Improvements in torque ripple will give additional benefit in losses and hence lesser load on thermal system. This study investigates the possible usage of PMSM and field-oriented control for future generation reaction wheels and its implementation in FPGA. Simulation results showed improved steady-state performance in terms of torque ripple without any degradation in the transient response of the wheels. The study can be well extended to gimbal control of CMG and other high-precision servo applications in the spacecraft.

Keywords PMSM · BLDC · SVPWM · FOC

1 Introduction

In BLDC motor with six-step commutation, current has to rise sharply through the stator-winding inductance. This leads to torque ripple, which prevents the usage of these motor for precise servo application [1, 2]. The space vector modulation technique improves the system performance with low torque ripple, thus making it suitable for high-precision applications employing electromechanical actuators [3]. In order to evaluate the current oscillation and torque ripple generated by BLDC

K. Ratheesh (✉) · S. Sreejith · T. R. Haridas
ISRO Inertial System Unit, Thiruvananthapuram 696013, India
e-mail: ratheesh_kooleri@vssc.gov.in

© The Author(s), under exclusive license to Springer Nature Singapore Pte Ltd. 2023

R.S. Priyadarshini and T. Sundararajan (eds.), *Advances in Small Satellite Technologies*, Lecture Notes in Mechanical Engineering, https://doi.org/10.1007/978-981-19-7474-8_13

motor, every instant of power device turning on and off is to be simulated. To make results correct, the input phase current is to be synchronized with the back EMF. Here, the primary focus is to control the motor in d-q axis using field-oriented control and space vector modulation techniques. Vector control technique will help to improve the torque ripple and efficiency, making it suitable for applications employing electromechanical actuators [4–6]. Simulation and modelling of permanent magnet brushless DC motor drive [7] gives a baseline modelling method for general BLDC motors. However, in this study a more detailed modelling approach specific to reaction wheels of Indian IRS satellites is carried out to investigate more realistic behaviour of BLDC motor. Digital implementation of FOC in FPGA is another challenge. The implementation of the permanent magnet synchronous motor (PMSM) controller by using field-oriented control (FOC) method has been discussed. A scheme for FOC using a fixed 32-bit DSP controller TMS320F2808 controller [8] was taken as a baseline and implemented using generic VHDL to avoid processor dependency. Earlier reports have discussed the design of space vector pulse width modulation (SVPWM) integrated circuit [9, 10] in FPGA and CORDIC algorithm to solve trigonometry, hyperbolic and exponential function on FPGA [11]. This provides efficient hardware utilization of FPGA resources.

Literature survey shows that even though various control strategies in controlling the brushless DC motor have been presented as above, the area of space vector PWM for controlling the brushless DC motor/PMSM has not been much investigated. Usage of vector-controlled PMSM for reaction wheel and CMG applications are not explored much. The comparative study of FOC control of BLDC and PMSM are not investigated. A complete simulation model as detailed in this study will help to optimize the performance of the reaction wheel assembly. As the torque ripple and micro vibration are closely related, improvement in the torque ripple performance will reduce the micro-vibrations in the spacecraft compared to existing scheme. All the sub-modules required for FOC has been implemented in view of optimized use of FPGA resources and tested in Actel fusion FPGA-AFS600.

2 Field Oriented Control (FOC) for PMSM Motors

In a BLDC/PMSM motor, torque produced can be maximized when stator mmf vector is in quadrature with the rotor flux vector. The main aim of FOC is to maximize the torque produced. In this scheme, three-phase current of the motor stator is measured in the stationary coordinate system of the stator which is further converted in to two-phase current in the stator coordinates. This transformation is known as Clarke transformation. Furthermore, this current is converted into the rotor coordinate system known as d-q reference frame of the rotor. This transformation is known as Park transformation. Two currents obtained by these transformations are direct and quadrature components. Direct component will not produce any torque. Quadrature component of current is responsible for torque. Current obtained will appear as DC upon which PI controller can be operated easily. Inverse Park and Clarke transformation are applied and corresponding reference vectors of space vector PWM are generated. To model the field produced by the winding currents in the stator, 'current space vectors' are used. The direction and magnitude of field produced in the winding is same as that of current space vector. The total stator field is the vector sum of three current space vector components. The d axis current (direct component) is set to zero so as to minimize any unwanted torque given to the motor, since the direct component produces torque that is aligned with the rotor.

$$V_a = V_m \times \sin(\omega t) \tag{1}$$

$$V_b = V_m \times \sin(\omega t - 120^\circ) \tag{2}$$

$$V_c = V_m \times \sin(\omega t + 120^\circ) \tag{3}$$

The above three vectors (1) can be combinedly represented by a single vector which is known as space vector and it is represented as,

$$V_s = V_a + V_b e^{\frac{j2\pi}{3}} + V_c e^{-\frac{j2\pi}{3}} \tag{4}$$

Following transformations are performed.

2.1 *Clarke Transform*

$$\begin{bmatrix} i_\alpha \\ i_\beta \end{bmatrix} = \begin{bmatrix} 1 & -\frac{1}{2} & -\frac{1}{2} \\ 0 & \frac{\sqrt{3}}{2} & -\frac{\sqrt{3}}{2} \end{bmatrix} \begin{bmatrix} i_a \\ i_b \\ i_c \end{bmatrix} \tag{5}$$

where,

i_a, i_b, i_c are the three-phase currents in the 1200 stationary reference frame of stator.
$i_{\alpha,} i_\beta$ are the two orthogonal currents in the ationary reference frame where i_α is aligned to i_a.

2.2 *Park Transformation*

$$\begin{bmatrix} i_d \\ i_q \end{bmatrix} = \begin{bmatrix} \cos\theta & \sin\theta \\ -\sin\theta & \cos\theta \end{bmatrix} \begin{bmatrix} i_\alpha \\ i_\beta \end{bmatrix} \tag{6}$$

where,

$i_{d,}i_q$ are two orthogonal currents in the rotor reference frame with i_d aligned along flux vector of a pole pair.
θ is the angle between the rotating and stationary reference frame.

2.3 Inverse Park Transform

$$\begin{bmatrix} V_\alpha \\ V_\beta \end{bmatrix} = \begin{bmatrix} \cos\theta & -\sin\theta \\ \sin\theta & \cos\theta \end{bmatrix} \begin{bmatrix} V_d \\ V_q \end{bmatrix} \tag{7}$$

where,

V_d, V_q are two orthogonal voltages proportional to outputs of PI controller in the rotor reference frame.
V_α, V_β are two orthogonal voltages in the stationary reference frame, with θ as the angle between the rotating and stationary reference frame.

The following equation represents a general transformation between three phases to two-phase voltages and resultant vector in stationary frame.

$$\begin{bmatrix} V_\alpha \\ V_\beta \end{bmatrix} = \begin{bmatrix} 1 & \frac{-1}{2} & \frac{-1}{2} \\ 0 & \frac{\sqrt{3}}{2} & \frac{-\sqrt{3}}{2} \end{bmatrix} \begin{bmatrix} V_a \\ V_b \\ V_c \end{bmatrix} V_{\text{ref}} = \sqrt{(V_\alpha)^2 + \left(V_\beta\right)^2} \theta = \tan^{-1}\frac{V_\alpha}{V_\beta} \tag{8}$$

The required space vector in SVPWM technique is generated by null vector and two adjacent vectors. The calculation of sector in which the voltage vector belongs can be obtained from Eq. (8). The sector number is used for timing signal generation and duty cycle calculation.

2.4 Duty Cycle Calculation

General equation for duty cycle in each sector is

$$T_{k+1} = 3T/Vdc\left(v_\alpha \sin\frac{k\pi}{3} - v_\beta \cos\frac{k\pi}{3}\right) \tag{9}$$

$$T_{k+1} = 3T/Vdc\left(v_\alpha \cos\frac{(k-1)\pi}{3} - v_\beta \sin\frac{(k-1)\pi}{3}\right) \tag{10}$$

where T_k and T_{k+1} are switching time for vectors k and $k + 1$, respectively, where k denotes sector number (1, 2, …6).

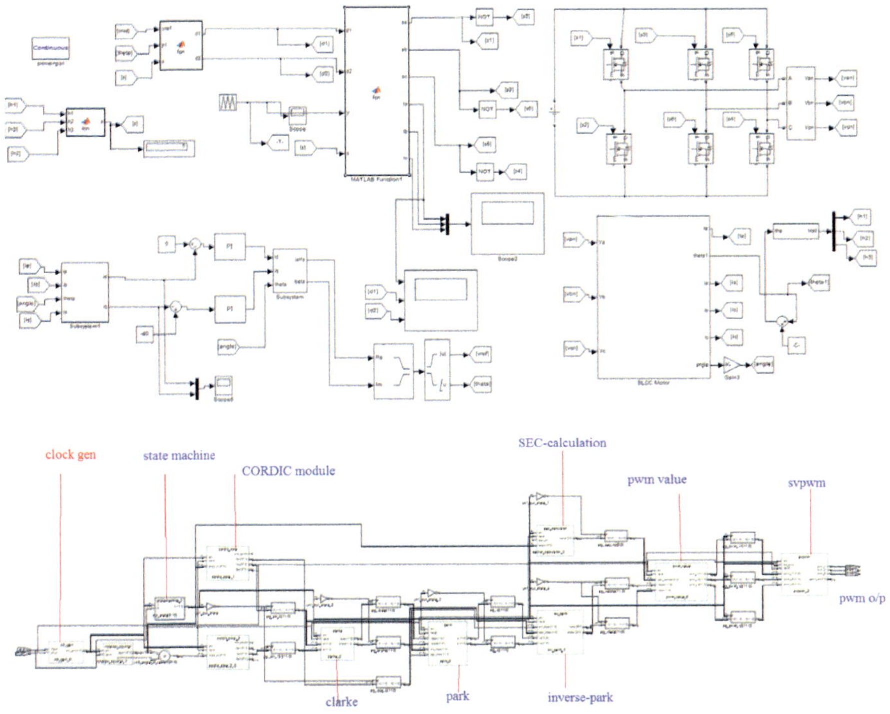

Fig. 1 Complete closed-loop Simulink model of field-oriented control for PMSM motor and hierarchical view of hardware developed inside Actel fusion FPGA-AFS600

The Fig. 1 shows the (a) complete closed loop model of field-oriented control for PMSM motor developed in MATLAB Simulink and (b) hierarchical view of main module developed in Actel fusion FPGA- AFS600.

3 Analysis

The stator-side equations for each of the phases are given as follows [3]:

$$\begin{bmatrix} V_a \\ V_b \\ V_c \end{bmatrix} = \begin{bmatrix} R & 0 & 0 \\ 0 & R & 0 \\ 0 & 0 & R \end{bmatrix} \begin{bmatrix} i_a \\ i_b \\ i_c \end{bmatrix} + \begin{bmatrix} L-M & 0 & 0 \\ 0 & L-M & 0 \\ 0 & 0 & L-M \end{bmatrix} \frac{d}{dt} \begin{bmatrix} i_a \\ i_b \\ i_c \end{bmatrix} + \begin{bmatrix} e_a \\ e_b \\ e_c \end{bmatrix} \tag{11}$$

V_a, V_b and V_c represent motor phase voltages; L inductance, R resistance, M mutual inductance, and e_a, e_b and e_c are the back EMFs of each of the phases.

The instantaneous back EMF generated in a 3-phase BLDC motor is a function of rotor position

$$e_a = K_b(\theta)\omega_m \tag{12}$$

$$e_b = K_b(\theta - 2\pi/3)\omega_m \tag{13}$$

$$e_c = K_b(\theta + 2\pi/3)\omega_m \tag{14}$$

where K_b is back EMF constant in V/(rad/sec) and ω_m speed of motor in rad/sec.

$$T_e = (e_a i_a + e_b i_b + e_c i_c)/\omega_m \tag{15}$$

The electromagnetic torque T_e is also related to motor constant, phase current and back EMF function which is given as

$$T_e = K_t(f_a(\theta)i_a + f_b(\theta)i_b + f_c(\theta)i_c) \tag{16}$$

where $f_a(\theta)$, $f_b(\theta)$, $f_c(\theta)$ are back EMF functions.

The developed torque T_e is a function of back EMF and stator current and hence the smoothness and close match between the two determines the torque ripple performance of the motor. In view of this, reaction wheels with FOC control and PMSM motor is expected to have lower torque ripple compared to the six-step commutation with BLDC motor or FOC scheme of BLDC. Six-step commutation considers that the commutation of phase current is immediate which is not the case as in practical systems due to rise time of current in motor coils. FOC control on BLDC although avoids sudden commutation changes, but has mismatch between the sinusoidal current and trapezoidal back EMF resulting in torque ripple.

The equation of flywheel motion in a motor is

$$J(d\ \omega_m/dt) + B\ \omega_m = T_e - T_l \tag{17}$$

where B represents coefficient of friction and J is the inertia of the rotor, which will be negligibly small and T_l is load torque.

In the PMSM motor, the back EMF function mentioned in Eq. (16) can be replaced by three 120 degree-shifted sinusoidal. PMSM/BLDC motor model has been developed in Simulink. The equations mentioned above have been used for modelling. The exact value of inertia, friction, torque constant and back EMF constant of reaction wheels and magnetic suspension reaction wheels in IRS satellites are used for modelling. FOC control modelled in MATLAB Simulink and simulation results were generated. It involves the modelling of Park and Clarke transformation and SVPWM.

4 Simulation Results and Implementation

The phase voltage obtained from the simulation model shows that switching between two space vectors happens during motoring. The phase voltages of all the three phases have been generated. It can be observed that the FOC control of BLDC also leads to a torque ripple of about 7% of the developed torque due to mismatch in trapezoidal back EMF and sinusoidal phase current. As the phasc current waveforms are smooth sinusoidal and closely matching to sinusoidal back EMF, torque ripple observed in FOC control of PMSM is very small (< 1%). A comparative study on the FOC model of BLDC, PMSM and trapezoidal control has been carried out to establish the advantage of FOC with PMSM motor. Torque ripple observed in trapezoidal control can be completely eliminated in FOC control of PMSM. Table 1 shows the comparison of simulation results.

After evaluating the performance of the FOC control by Simulink model, VHDL code has been developed and implemented in ACTEL fusion FPGA AFS600. Implementations of different sub-modules are carried out in view of optimizing the resources of the target FPGA. It used only 56% core cells for implementing entire algorithm. The fused FPGA is tested by applying appropriate input to the required I/O pins of the FPGA. The output SVPWM waveforms were captured by using digital oscilloscope. The sub-module such as CORDIC, SVPWM, PWM value generation, sector calculation, Clarke transformation, Park transformation and inverse Park transformation are implemented in VHDL.

Park transformation is implemented in VHDL to generate a DC value from two orthogonal I-alpha and I-beta signals of Clarke module. That is, here the two-phase orthogonal vectors laying in the stationary frame are transformed into two orthogonal vectors laying in the rotating frame of the rotor.

Here, from the inverse Park output, SVPWM module will calculate duty cycle value required for each sector vectors. After that, this value will be converted into a form compatible to PWM generation, i.e. the vectors are to be generated in such a way that minimum switching should happen for the driver circuits. Here, at each sector, resultant is decided by two vectors and a null vector. Timing signal for each phase is generated by this section.

The resource requirement (logic cells) for the FOC algorithm was highly reduced by the usage of CORDIC algorithm for sin and cosine generation. Algorithm was developed which was based on state machine structure, iterative and pipelined. A 27-bit CORDIC algorithm was developed to eliminate the truncation error as 24 bits

Table 1 Comparison of simulation results

Method in reaction wheel	Developed torque (Nm)	Torque ripple (%)
Trapezoidal control with PWM (existing)-BLDC based	0.25–0.3	< 9.09
Field oriented control BLDC	0.225–0.26	< 7.21
Field oriented control PMSM	0.2	< 0.5

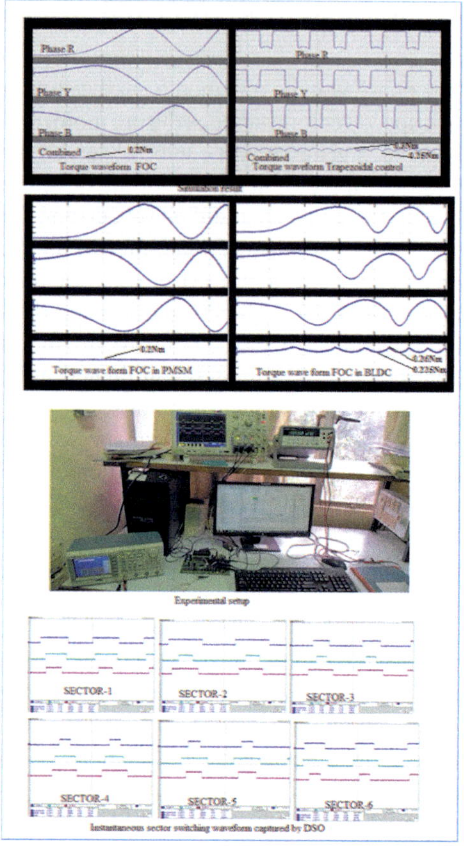

Fig. 2 Simulation results (torque in Nm at vs. time in milli seconds), experimental setup and hardware level sector waveforms (volt vs. time in milliseconds)

are used for other modules. The developed code can be used for numerical controlled oscillator, whose frequency and phase can be controlled. The VHDL implementation was used to drive a three-phase load using a H-bridge power stage to verify the sector switching and phase waveforms.

Figure 2 shows simulation results, experimental setup and hardware level sector waveforms captured in FPGA pin using storage oscilloscope. It clearly shows the wave form obtained for different sectors.

5 Conclusions

The simulation results reveal that the PMSM and field-oriented control (FOC)-based reaction wheels can give improved performance with respect to torque ripple over the trapezoidal drive-based BLDC reaction wheels used in Indian remote sensing space crafts.

A resource-optimized generic hardware description language implementation of the SVPWM, CORDIC algorithm, and forward-inverse Park and Clarke transform was carried out and tested using fusion FPGA kit (AFS600).

The study can be extended in applications such as gimbal control of control momentum gyroscopes and other high-precision servo applications in the spacecraft.

Acknowledgements We are thankful to Dr Sam Dayala Dev, Director ISRO Inertial System Unit (IISU), for providing laboratory and knowledge resource facilities. We take this opportunity to convey our deep sense of gratitude to all the team members of Satellite Systems Group (SSG) of ISRO Inertial System Unit without whom it would have been impossible to complete this study.

References

1. Pillay P, Krishnan R (1989) Modeling, simulation, and analysis of permanent-magnet motor drives, part II&I: the brushless dc motor drive. IEEE Trans Ind Appl IA-25(2):274–279
2. Lu H, Zhang L, Qu W (2006) A new torque control method for torque ripple minimization of BLDC motors with un-ideal back EMF. IEEE Trans Power Electron 21(6):1762–1768
3. Viswanathan V, Jeevananthan S (2011) A novel current controlled space vector modulation based control scheme for reducing torque ripple in brushless DC drives. Int J Comput Appl 28(2):25–31
4. Li WH, Chen Z-Y, Cao W-P (2012) Simulation research on optimization of permanent magnet synchronous motor sensor less vector control based on MRAS. In: IEEE international conference wavelet active media technology and information processing (ICWAMTIP)
5. John JP, Kumar SS, Jaya B (2011) Space vector modulation based field oriented control scheme for brushless DC motors. In: IEEE international conference, emerging trends in electrical and computer technology (ICETECT)
6. Liu Y, Zhu ZQ, Howe D (2007) Commutation-Torque-ripple minimization in direct-torque-control PM brushless DC drives. IEEE Trans Ind Appl 43:1012–1017
7. Tao GL, Ma ZY, Zhou LB, Li ZB (2004) Modeling simulation and experiment of permanent magnet brushless DC motor drive. In: 39th International universities power engineering conference, UPEC
8. Samar, A, Saedin P, Tajudin AI, Adni N (2012) The implementation of field oriented control for PMSM drive based on TMS320F2808 DSP controller. In: IEEE international conference on control system, computing and engineering (ICCSCE)
9. Yuan Z, Fei-peng X, Zhao-yong Z (2006) Realization of an FPGA-based space-vector PWM controller. In: IPEMC CES/IEEE 5th international, power electronics and motion control conference, vol 3
10. Nicola CI, Nicola M, Vintilă V (2019) Identification and sensorless control of PMSM using FOC strategy and implementation in embedded system. In: International conference on electromechanical and power systems (SIELMEN)
11. Bhuria S, Muralidhar P (2010) FPGA implementation of sine and cosine value generators using CORDIC algorithm for satellite attitude determination and calculators. In: 2010 International conference power, control and embedded systems (ICPCES)

Mechanical System Design for Small Satellites

Structural Analyses for a Typical Small Satellite

G. Biju, T. Sundararajan, and S. Geetha

Abstract Miniaturisation of electronics, newer materials especially composites and advanced manufacturing techniques including 3D printing has encouraged the development of small satellites. This enabled many commercial companies to venture in to this area since the development can be made possible with small budget. The market size of small satellite is expected to flourish in the next few years. Small satellites are used mainly in the areas of Earth observation services, monitoring of agricultural fields, detection of climatic changes, disaster mitigation, meteorology, etc. In the current day scenario of launch vehicle industry, every mission of a launch vehicle is unique since they are used for putting satellites of different mass, size, shape, purpose and to various orbits. Satellites are exposed to dynamic/fluctuating excitations due to different dynamic environment from lift off to the completion of its intended mission. The dynamic environment of the satellite will be different for different launch vehicle configuration and its mounting scheme. This paper addresses quasi-static and dynamic analyses carried out for a typical small satellite using indigenously developed finite element software FEAST^{SMT} by Vikram Sarabhai Space Centre (VSSC). This study helps in assessing the structural adequacy of small satellite as well as the acceleration level experienced by different components and packages. This will help the satellite team to modify the structure, change the mounting location or by adopting new mounting schemes of components/packages in order to optimise the design well before realising the hardware.

Keywords FEAST^{SMT} · Small satellite · Frequency response · Random response

1 Introduction

Small satellite is one of the fastest growing segments in space applications. The miniaturisation of electronics, newer and newer materials especially composites as well as manufacturing including 3D printing has encouraged the development of

G. Biju (✉) · T. Sundararajan · S. Geetha
Vikram Sarabhai Space Centre, ISRO, Thiruvananthapuram, India
e-mail: bijutkmce@vssc.gov.in

© The Author(s), under exclusive license to Springer Nature Singapore Pte Ltd. 2023

R.S. Priyadarshini and T. Sundararajan (eds.), *Advances in Small Satellite Technologies*, Lecture Notes in Mechanical Engineering, https://doi.org/10.1007/978-981-19-7474-8_14

small satellites. Usually, the small satellites are launched along with main satellite as co-passengers. These co-passengers are mostly accommodated on equipment bay (EB) structure or on multi-satellite adaptors through interface ring and appropriate separation system. Launch vehicle and spacecraft are exposed to dynamic/fluctuating excitations due to different dynamic environment [1] from lift off to the completion of its intended mission. Different dynamic environments are wind and turbulence, motor generated acoustic noise, engine cut off and ignition thrust transients, aerodynamic forces, solid motor pressure oscillations, solid motor generated pressure spikes, stage separation, payload fairing separation, POGO, etc. These dynamic environments create random, periodic or transient excitation of different frequency regimes to vehicle and spacecraft. The acceleration level experienced by satellite at the interface with launch vehicle is used to carry out qualification/acceptance test and analysis. This is one of the several important analyses carried out to ensure that the mission objectives can be achieved without any mechanical damage to the satellite by launch vehicle vibration environments. The satellite finite element model will be subjected to these quasi-static, sine and random levels to estimate the stress and acceleration response of all packages to check the design adequacy.

In this paper, a small satellite of size 350 mm $\times$ 350 mm $\times$ 450 mm and mass nearly 35 kg is analysed using indigenous finite element software FEAST^{SMT} [2]. Detailed 3D model of the satellite is made using the pre-processor PREWIN. The main structure of satellite is made of sandwich panel of 30 mm core with 0.5 mm aluminium face sheets. All panels are joined by edge splicers. The satellite is mounted to aluminium interface ring of 60 mm height. Analysis [3] is carried out to estimate the stress due to quasi-static acceleration, first few global frequencies, its modal effective mass and response at important locations due to sine and random vibration. The sine and random levels considered in this analysis are the envelope of all dynamics events launch vehicle faces during the mission.

As part of indigenization process and to build competency in the area of computational numerical simulation for aerospace applications in ISRO, development of finite element-based structural analysis software, FEAST^{SMT} (Finite Element Analysis of STructures) was initiated. Over the years, the software has attained a greater level of maturity in the area of structural engineering analysis and catered most of the requirements of ISRO. The current version of the software includes analysis capabilities such as linear static, viscoelastic, free-vibration, buckling, transient, frequency and random response. These capabilities are supported by a rich element library comprising of beam, shell, solid, axisymmetric, spring, rigid-links, gap, etc. and can handle isotropic, orthotropic, incompressible and layered orthotropic material models. It is designed to use the hardware resources optimally, and hence, it can run in a range of platforms from personal computers to parallel processing systems.

2 Finite Element Model and Analysis

The finite element model used to simulate the satellite is shown in Fig. 1. The panels of the satellite were modelled by laminated sandwich shell elements. The package masses are lumped at appropriate location on the panels. The satellite is attached to interface ring which is modelled using shell element and fixed boundary condition is simulated at aft end of this interface ring.

2.1 *Quasi-Static Analysis*

The satellite is subjected to accelerations during the entire launching phase causing tension and compression of the structure in axial and lateral directions. The maximum quasi-static acceleration (static + low frequency vibrations) is ± 11 g in longitudinal direction and ± 6 g in lateral directions. The stresses are estimated due to the body forces generated by these acceleration levels. The maximum normal stress on the honey comb panel facing sheet for axial quasi-steady load case is 0.33 MPa. Maximum shear stresses on the core for this load case are 1.31e-03 and 2.23e-03 MPa. Maximum normal stress of 0.17 MPa is obtained on the facing sheet during lateral quasi-static load applied in lateral (X) direction as shown in Fig. 2. Maximum shear stresses on core for same load case are 5.73e-04 and 1.27e-03 MPa.

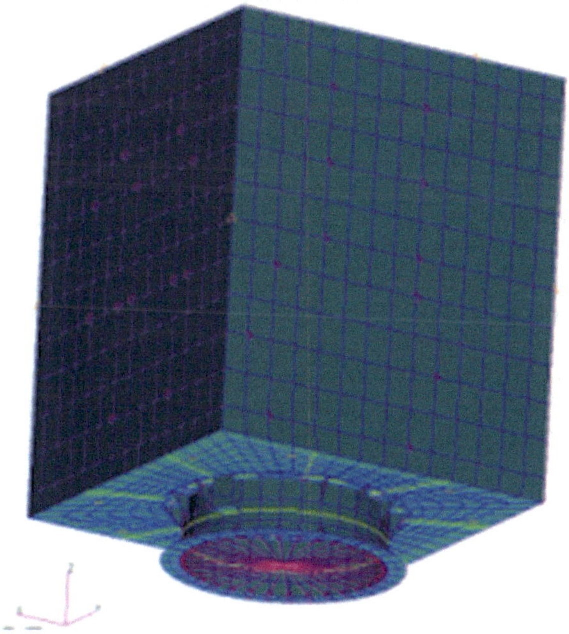

Fig. 1 Finite element model of a small satellite

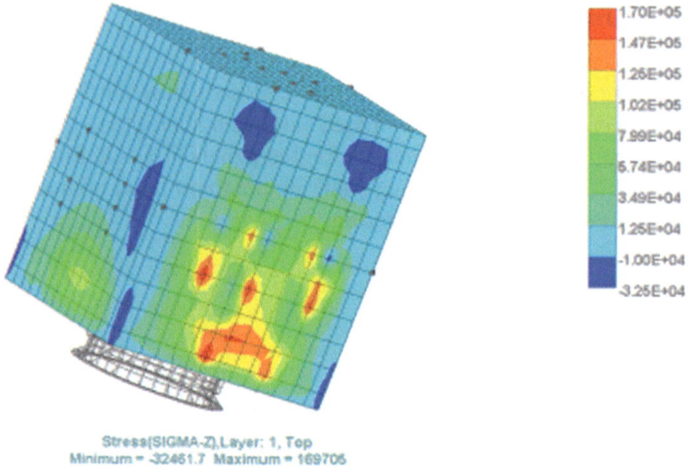

Fig. 2 Distribution of normal stress (σz) for lateral (X) quasi-steady load case

Usually sine burst (SB) excitation test is used for strength evaluation of satellite hardware as an alternate to static pull and centrifuge tests. The shaker controller is used to apply a sinusoidal base drive acceleration to the item at a pre-resonant or as close as possible to the rigid body mode, usually being not more than 1/3 of the first mode frequency of the test article. Because the test article is being accelerated as a rigid body, a uniform inertial body load is generated in the test article.

2.2 *Dynamic Analysis*

Structural dynamic parameters of the launch vehicle such as frequency, mode shape and damping are very important for mission critical studies such as digital auto pilot design, POGO stability studies as well as the dynamic envelope studies. The dynamic envelop is the usable volume for a spacecraft inside the payload fairing which very much depends on the satellite dynamic deflections. The specification for the lateral and axial frequencies of satellites is arrived considering the significant lateral and axial modes of the launch vehicle. Usually, it is kept two times that of launch vehicle significant modes. Table 1 gives the first lateral and axial frequencies and associated effective mass. Nearly, 50% of the total mass of the satellite is participating in these modes. Figure 3 shows the finite element model and mode shapes for first lateral and axial mode.

Table 1 Base fixed modes of the satellite

No	Frequency (Hz)	Modal effective mass (kg.)	Remarks
1	71.1	16.20	Lateral mode (ZY)
2	75.1	16.35	Lateral mode (ZX)
3	182.4	16.00	Axial mode

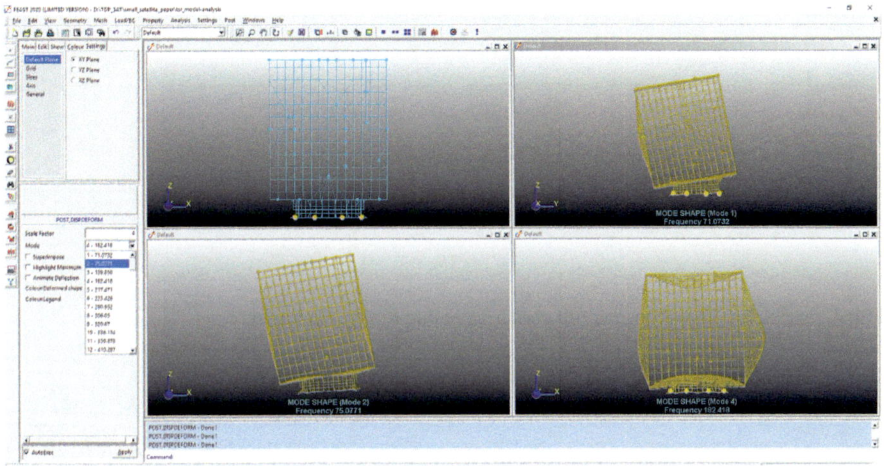

Fig. 3 PREWIN-pre/post window showing FE model and lateral and axial mode shapes of a typical satellite

2.3 *Response Analysis*

Response analysis of this typical small satellite is carried out in both sine and random environment. Sine test qualification level specified are 4.5 g (0 to peak) from 8 to 100 Hz for axial test and 3 g from 8 to 100 Hz for lateral axis. For random excitation [4], the PSD at different bands are from 20 to 110 Hz 0.002 g^2/Hz, from 110 to 250 Hz the level changes from 0.002 to 0.034 g^2/Hz, from 250 to 1000 Hz the level remains as 0.034 g^2/Hz and from 1000 to 2000 Hz the level changes from 0.034 to 0.009 g^2/Hz. Analysis is carried out to estimate accelerations at different package location as well on certain location on the structure. These excitations are given as base excitation as in test. Damping used in the analysis is 2% for all frequencies. Figure 4 shows the acceleration response at top corner node of the satellite structure due to sine excitation. The peak is at global lateral frequency 75.1 Hz, and the amplification factor is 14. Figure 5 shows the response at top corner node of the structure due to random excitation in lateral direction. Similar responses obtained from this analysis corresponding to component and package mounting locations has to be considered for its qualification.

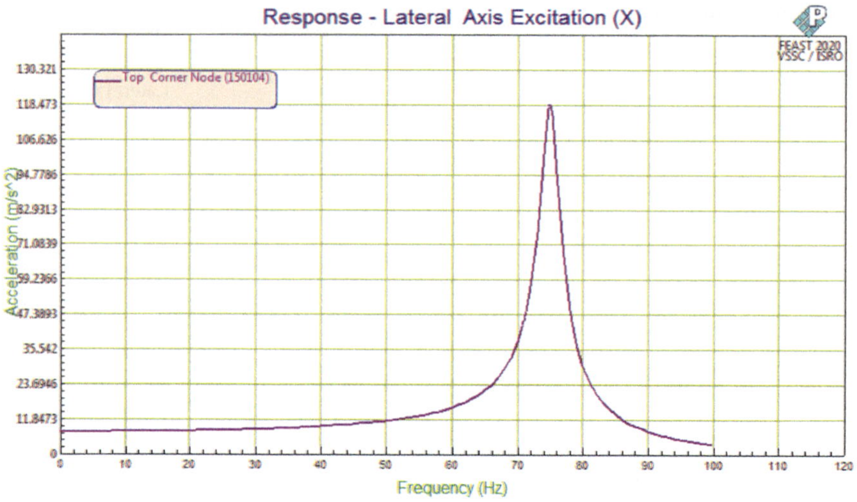

Fig. 4 Acceleration response at top corner node due to sine excitation

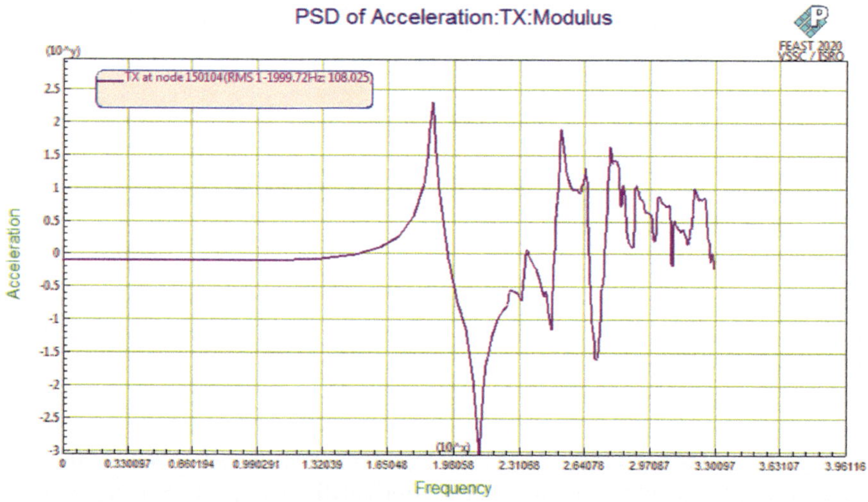

Fig. 5 Acceleration response at top corner node due to random excitation

3 Conclusion

This paper gives the details of quasi-static, dynamic characterisation and response analyses to be carried out for a typical small satellite design. The stress and the responses help the satellite team to assess the structural adequacy and vibration levels of the components and electronic packages during its ride with the launch vehicle. This analysis helps in modifying the structure or adopting alternate mounting schemes for packages/modules if required. The response estimated from these analyses can be used for qualification of components and packages mounted on to it.

References

1. Dynamic Environmental Criteria. NASA-HDBK-7005
2. Feast User's Manual. VSSC, ISRO
3. Abdelal GF, Abuelfoutouh N, Gad AH, Finite element analysis of satellite structures application to their design. Springer
4. Jaap W, Random vibrations in spacecraft structures design. Springer

Modeling Deployment of Tape Spring Antennas and Its Effects on CubeSat Dynamics

Aniruddha Ravindra Ranade and Salil S. Kulkarni

Abstract Tape spring mechanisms have been widely used to design deployable structures for satellites. In this paper, the dynamics of the deployment of VHF tape spring antennas from a CubeSat have been discussed. A set of nonlinear equations of motion representing the coupled dynamics of the deployment and the overall motion of the CubeSat were obtained using Kane's method. These were integrated numerically to obtain parameters of interest like the time taken for deployment, the trajectory of the antenna, and variation of the CubeSat angular velocities during the interval of deployment. These models have been used to analyze the design of Advitiy (1U CubeSat conceptualized by IIT Bombay's Student Satellite Team). Based on simulations, comments have been made on this design of the antenna deployment system, and few performance specifications have been laid down for the CubeSat.

Keywords Tape spring · Deployable antenna · Kane's method · CubeSat dynamics

1 Introduction

Over the last decade, rapid growth has been seen in the CubeSat market. Due to the miniaturization of technology, it is possible to send numerous small satellites or CubeSats rather than one "mega" satellite. Several deployable members like antennas and solar panels are typically incorporated in CubeSats, allowing a smaller volume during the launch. These members are deployed after reaching the desired orbit. The design of such mechanisms is a complicated task. Investigations to develop passive and active control for primitive spacecraft, which were typically rigid, have been carried out. The same analysis cannot be directly extended to modern spacecraft, the primary reason being the inclusion of several flexible members and deployable structures [5].

A. R. Ranade (✉) · S. S. Kulkarni
Department of Mechanical Engineering, Indian Institute of Technology, Bombay, Mumbai 400076, India
e-mail: aniruddha.ranade@iitb.ac.in

© The Author(s), under exclusive license to Springer Nature Singapore Pte Ltd. 2023
R.S. Priyadarshini and T. Sundararajan (eds.), *Advances in Small Satellite Technologies*, Lecture Notes in Mechanical Engineering, https://doi.org/10.1007/978-981-19-7474-8_15

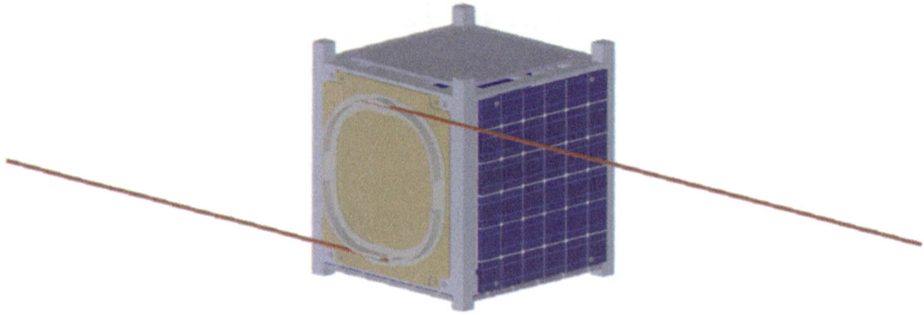

Fig. 1 Advitiy

This paper analyzes the antenna deployment system for CubeSats being developed by the Student Satellite Team, IIT Bombay.[1] The dynamics of the deployment are of interest as it influences several mission parameters. It is essential to know at what angular rates of the satellite can the antennas be deployed. Time taken for the deployment to complete is of interest. Based on the analysis, some remarks have been made on the design of other subsystems.

2 Design of the Antenna Deployment System for CubeSat Application

Student Satellite Team, IIT Bombay, has been developing an antenna deployment system for CubeSat applications [6]. This system comprises two tape spring antennas wrapped around a circular ring supported by a PCB. Advitiy, the second student satellite of IIT Bombay, was being designed with this antenna deployment system [4]. In Fig. 1, Advitiy, along with the antenna deployment system, is shown. Advitiy, a 1U CubeSat, was being designed for a Low Earth Orbit (LEO) mission. The amateur band 144–146 MHz was selected for communication. The monopole or dipole antennas for this band cannot be accommodated inside 1U volume constraints. To meet this, an antenna deployment system was designed. An antenna deployment system, capable of deploying dipole antennas of 51 cm length, was designed and prototyped by the team.

[1] Student Satellite Program, IIT Bombay, is an initiative taken up by the students to build a centre of excellence in space science and technology at IIT Bombay. The first satellite, Pratham, under this program, was launched onboard the PSLV-C35 on 26th September 2016. Information about current projects can be found here: https://www.aero.iitb.ac.in/satlab/.

3 Dynamics of the CubeSat Over the Interval of Deployment

Tape springs are straight, thin-walled strips with a curved cross-section [10]. A carpenter's tape is a tape spring. Due to its curvatures, it exhibits unique behavior. It is stored as a coil but can be extended into a stiff tape that can be easily folded. Except for the region of the fold, the rest of the tape spring is mostly straight and undeformed. Tape springs store elastic strain energy during folding and, in principle, would deploy freely into the straight unstrained configuration once the restraining forces have been released. Tape spring can be folded in two ways, depending upon the sense of curvatures. Several deployable structures that make use of tape springs have been analyzed [2, 9, 11]. Different modeling approaches have been proposed, ranging from a theoretical model [10] to develop a FE-based model [3, 7]. In this paper, a suitable extension is proposed to the theoretical model proposed by Seffen [10].

A model to study the dynamics of uncoiling of a tape spring mounted on a spool and its experimental verification is presented by Seffen [10]. This analysis is applicable when the radius of curvature of the tape spring and tape spool is nearly the same. The tape spring is modeled as a rigid body of varying length, which uncoils from the circular base. This modeling approach assumes the conservation of energy.

The antenna deployment system consists of a disk around which the antennas are wrapped. The CubeSat can be assumed to consist of a rigid cube with two flexible tape spring antennas wrapped around a disk mounted on one of the faces. Figure 2 shows the schematic and parameters used to denote the unwrapping of the antenna from the deployment system. The x and y axes are parallel with the face containing the antenna deployment system, and the z-axis is normal to the face with the antenna deployment system, pointing towards the reader. The x-axis is parallel to the deployed antenna and points towards the right side. It was assumed that no external torques and forces are acting on the system during the deployment. It is expected that the angular momentum of the system remains conserved. The system gains kinetic energy at the expense of the potential energy stored in the coiled tape spring.

Fig. 2 Schematic of the model

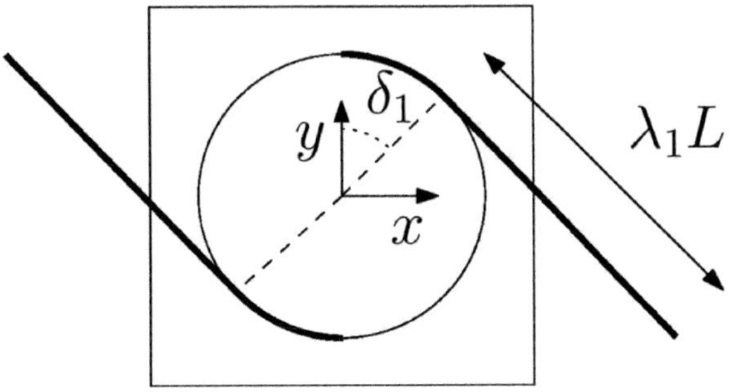

Table 1 Results of simulation on disk

Test ID	Parameters				Results	
	No. of antenna	Thickness (mm)	Direction of folding	Total energy (J)	Deployment time (sec)	Angular velocity (rad/sec)
FT01	1	0.2	Opposite	0.4115	0.177	−10.76
FT02	1	0.1	Opposite	0.0514	0.364	−2.86
FT03	1	0.1	Equal	0.0277	0.496	−2.10
FT04	1	0.2	Equal	0.2216	0.242	−7.92
FT05	2	0.2	Opposite	0.8231	0.170	−19.53
FT06	2	0.1	Opposite	0.1029	0.355	−5.40
FT07	2	0.1	Equal	0.0554	0.484	−3.96
FT08	2	0.2	Equal	0.4432	0.232	−14.35

Before developing the full model for the CubeSat, a more straightforward problem was analyzed - a disk with one or two antennas in a plane, where translation in two axes and rotation about one axis were permitted.[2] Kane's method was used to obtain the dynamics equations. This set of nonlinear differential equations was then integrated numerically. A set of simulations were performed to understand the effect of the number of antennas, thickness, initial energy, and the sense of folding for the tape springs on the final angular velocity attained and the time taken for the deployment. This was later extended for the CubeSat by simulating a cube with one or two antennas, where translations in three axes and rotations about three axes were permitted [8]. Values assigned for various parameters were obtained from the design of Advitiy.

4 Results

4.1 Simulations on Disk

Different test cases, their parameters, and the results have been tabulated in Table 1. For all these simulations, initially, the disk is at rest, and 0.99 fraction of the antenna is wrapped around the disk. The numerical integration is performed over two intervals. Timestep of 0.00001 was used from 0 s to 0.001 s, and a timestep of 0.001 was used from 0.001 s for the rest of the period of analysis. The mass of the disk is the same as that of the satellite, and the moment of inertia for the disk is the same as that for the satellite, about the z-axis.

[2] Refer Appendix—Derivation of dynamics of disk with 1 coiled antenna.

Table 2 Results of simulations on cube

Test ID	Parameters				Results	
	No. of antenna	Thickness (mm)	Direction of folding	Total energy (J)	Deployment time (sec)	Angular velocity (rad/sec)
FT09	1	0.2	Opposite	0.4115	0.177	{3.98, 1.08, – 11.59}
FT10	2	0.2	Opposite	0.8236	0.173	{4.04, 1.65, – 20.13}

4.2 Simulations on Cube

Different test cases, their parameters, and the results have been tabulated in Table 2. For all these simulations, initially, the cube is at rest, and 0.99 fraction of the antenna is wrapped around. The numerical integration is performed over two intervals. Timestep of 0.00002 was used from 0 to 0.001 s, and a timestep of 0.001 for FT09 and 0.002 for FT10 was used from 0.001 s for the rest of the period of analysis.

For the simulations performed, the velocity of the center of mass and the angular momentum were monitored. It was observed that the velocity of the center of mass remains nearly zero, and the angular momentum is conserved. These slight variations have been attributed to numerical errors, but they are within engineering approximations and can be neglected. Since the system begins from rest, the total energy is equal to the potential energy stored due to the tape spring coiling. This potential energy depends on the strain energy per unit area and the coiled area of the tape. The opposite sense of folding is responsible for a higher value of the strain energy, while the equal sense of folding is responsible for a lower value. A thicker antenna, when coiled, stores more strain energy than a thinner antenna.

The plots for test case FT10 are shown. Figure 3 denotes the plot of the fraction of deployment for both the antennas over time. Both the antennas take nearly the same amount of time for deployment. Figure 4 shows the angular velocity of the cube. ω_1, ω_2, and ω_3 denote angular rates about the 3 axes, where ω_3 is about the z-axis, about which the antennas are wrapped around. Hence, maximum change is observed about the z-axis. The non-zero cross-terms in the moment of the inertia matrix are responsible for angular rates about the other two axes.

5 Conclusions

In this paper, a model for coupled dynamics of the deployment of the tape spring antennas mounted on disk and cube were derived, and simulations were performed. The following qualitative remarks can be made from the analysis:

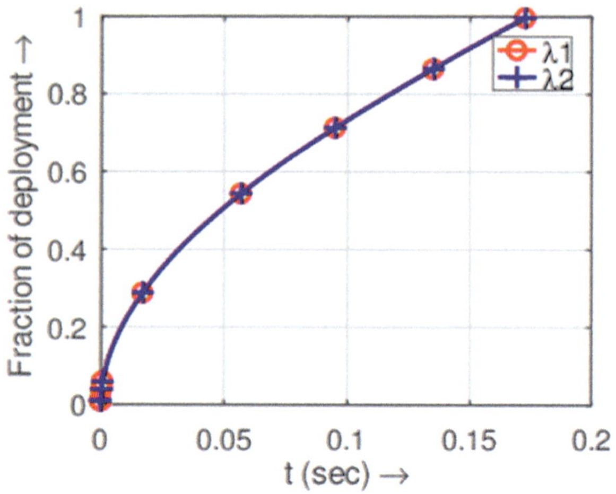

Fig. 3 Fraction of deployment

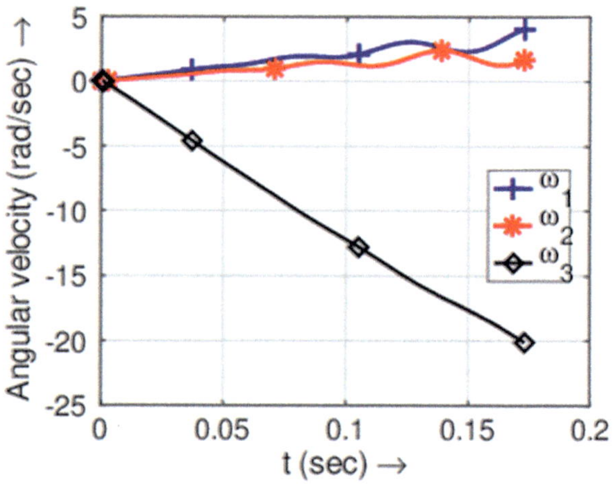

Fig. 4 Angular rates of CubeSat

- The maximum change in angular velocity is observed along the axis about which the antennas are wrapped around.
- A thicker antenna with an opposite sense of bending produces the maximum angular velocity, while a thinner antenna with an equal sense of bending produces the least angular velocity.
- For the same thickness and direction of folding, the system with one antenna and two antennas have nearly the same deployment time, with the system with two antennas requiring slightly less time than the system with one antenna.

Following are some considerations which shall influence the design of other subsystems:

- The initial angular velocity of Advitiy, while in orbit, may be non-zero. Depending on the direction of the unfolding of the antennas, this can either increase or decrease the angular velocity. Hence, the onboard logic should check the angular rates and factor in the direction of the unfolding of antennas. Based on this, an appropriate instant should be decided to issue the command for deployment.
- From the model, the anticipated angular rates with which Advitiy spins after deployment can be evaluated. The attitude determination and control subsystem should be able to control Advitiy under these conditions. Thus, a set of requirements must be generated by performing a more exhaustive set of simulations on the model.

Acknowledgements We want to thank the Student Satellite Team for their inputs, which made this research work possible, and the Department of Mechanical Engineering for their support.

Appendix—Derivation of Dynamics of Disk with 1 Coiled Antenna

The system with a uniform disk and one tape spring antenna wrapped around its periphery is considered. The schematic is shown in Fig. 5. The disk has radius r, mass m_d, and moment of inertia about its center of mass I_d. The tape spring has length L, mass per unit length ρ, the radius of curvature R, and thickness t. The inertial frame is defined using N. The frame attached to the disk, located at its center of mass, is denoted by A. At $\psi = 0$, the frames N and A are parallel to each other. x and y denote the position of the center of mass of the disk in N frame. Let λ_1 denote the fraction of length of tape spring, which has been unwrapped. Let δ_1 denote the angle subtended at the center of the disk by the portion of tape spring coiled, at a given instant. The subscript has been added to identify the antenna. There are two parts of the tape spring, a coiled section, and an unwrapped section. A generic point on the coiled section is denoted by **xP11Aa** and on the unwrapped section by **xP12Aa.** ζ_1 is the parameter, which is dependent on δ_1. Kane's method has been used to derive these equations [1, 8]. **xANn** denotes the position vector of the center of mass of disk with respect to frame N, expressed in unit vectors of N frame. Similar notations have been used throughout.

$$\mathbf{vANa} = \{u_1,\ u_2, 0\}^T \tag{1}$$

$$\mathbf{\&ANa} = \{0, 0, u_3\}^T \tag{2}$$

u_1, u_2, u_3 are generalized speeds.

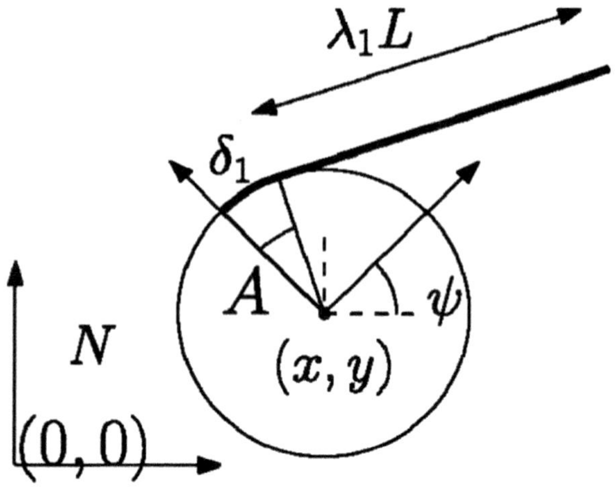

Fig. 5 Schematic for 1 antenna uncoiling from disk

$$L(1-\lambda_1) = r\delta_1 \tag{3}$$

$$\mathbf{xP11Aa} = \{r\ \sin(\zeta_1 L/r),\ r\ \cos(\zeta_1 L/r),\ 0\}^T \tag{4}$$

$$\mathbf{xP12Aa} = \{r\ \sin(\delta_1) + (L\ \zeta_1 - r\ \delta_1)\ \cos(\delta_1),\ r\cos(\delta_1) \\ -(L\ \zeta_1 - r\ \delta_1)\ \sin\ (\delta_1),\ 0\}^T \tag{5}$$

Velocities are written as follows. $\mathrm{d}\delta_1/\mathrm{d}t$ is set as u_4, the 4th generalized speed, in addition to the 3, defined before.

$$\mathbf{vP11Na} = \mathbf{vANa} + \mathrm{d}\mathbf{xP11Aa}/\mathrm{d}t + \mathbf{\&ANa} \times \mathbf{xP11Aa} \tag{6}$$

$$\mathbf{vP12Na} = \mathbf{vANa} + \mathrm{d}\mathbf{xP12Aa}/\mathrm{d}t + \mathbf{\&ANa} \times \mathbf{xP12Aa} \tag{7}$$

Using these, partial velocities (i =4 terms each. p indicates that these are partial velocities)-$\mathbf{vANap_i}$, $\mathbf{vP11Nap_i}$, $\mathbf{vP12Nap_i}$, and partial angular velocities $\boldsymbol{\omega}\mathbf{ANap_i}$ are evaluated by taking derivatives with generalized speeds. Accelerations are evaluated as follows.

$$\mathbf{aANa} = \mathrm{d}\mathbf{vANA}/\mathrm{d}t + \mathbf{\&ANa} \times \mathbf{vANa} \tag{8}$$

$$\mathbf{aANa} = \mathrm{d}\mathbf{\&ANa}/\mathrm{d}t \tag{9}$$

$$\mathbf{aP11Na} = \mathrm{d}\mathbf{vP11Na}/\mathrm{d}t + \mathbf{\&ANa} \times \mathbf{vP11Na} \tag{10}$$

$$\mathbf{aP12Na} = \mathrm{d}\mathbf{vP12Na}/\mathrm{d}t + \mathbf{\&ANa} \times \mathbf{vP12Na} \tag{11}$$

The potential energy for the coiled section of the tape spring is written as PE = $\mu R^2 \alpha \delta_1$ where μ denotes strain energy per unit area, α angle of the embrace of the cross-section. Generalized inertia forces, $F^*{}_i$, and generalized active force, F_i, are written

$$\begin{aligned} F_i^* = -\Big[& m_d \mathbf{aANa} \cdot \mathbf{vANap_i} + I_d \mathbf{aANa} \cdot \mathbf{\&ANap_i} \\ &+ \int_0^{r\,\delta 1\,/L} \rho L(\mathbf{aP11Na} \cdot \mathbf{vP11Nap_i})\mathrm{d}\,\zeta_1 \\ &+ \int_{r\,\delta 1\,/L}^{1} \rho L(\mathbf{aP12Na} \cdot \mathbf{vP12Nap_i}) d\,\zeta_1 \Big] \end{aligned} \tag{12}$$

$$F_4 = -\mathrm{dPE}/\mathrm{d}\delta_1 \tag{13}$$

Other generalized active force terms are zero. Using these, equations of motion are obtained.

$$\begin{aligned} &-(L\rho + m_d)\mathrm{d}u_1/\mathrm{d}t \\ &-0.5\rho\big[-2r^2\sin(\delta_1) - 2r(L - r\,\delta_1)\cos(\delta_1) + (L - r\,\delta_1)^2\sin(\delta_1)\big]\mathrm{d}u_3/\mathrm{d}t \\ &+0.5\rho\big[(L - r\,\delta_1)^2\sin(\delta_1)\big]\,\mathrm{d}u_4/dt = -(L\rho + m_d)u_2u_3 \\ &+2r^2u_3^2(1 - \cos(\delta_1)) + 2r(L - r\,\delta_1)\big(u_4^2 - u_3^2\big)\sin(\delta_1) \\ &+(L - r\,\delta_1)^2(u_4 - u_3)^2\cos(\delta_1) \end{aligned} \tag{14}$$

$$\begin{aligned} &-(L\rho + m_d)\mathrm{d}u_2/\mathrm{d}t \\ &-0.5\rho\big[2r^2(1 - \cos(\delta_1)) + 2r(L - r\,\delta_1)\sin(\delta_1) + (L - r\,\delta_1)^2\cos(\delta_1)\big]\mathrm{d}u_3/\mathrm{d}t \\ &+0.5\rho\big[(L - r\,\delta_1)^2\cos(\delta_1)\big]\,\mathrm{d}u_4/\mathrm{d}t = (L\rho + m_d)u_1u_3 - 2r^2u_3^2\sin(\delta_1) \\ &+2r(L - r\,\delta_1)\big(u_4^2 - u_3^2\big)\cos(\delta_1) + (L - r\,\delta_1)^2(u_4 - u_3)^2\sin(\delta_1) \end{aligned} \tag{15}$$

$$\begin{aligned} &0.5\rho\big[-2r^2\sin(\delta_1) - 2r(L - r\,\delta_1)\cos(\delta_1) + (L - r\,\delta_1)^2\sin(\delta_1)\big]\mathrm{d}u_1/\mathrm{d}t \\ &+0.5\rho\big[2r^2(1 - \cos(\delta_1)) + 2r(L - r\,\delta_1)\sin(\delta_1) + (L - r\,\delta_1)^2\cos(\delta_1)\big]\mathrm{d}u_2/\mathrm{d}t \\ &+\big[I_d + \rho L r^2 + (\rho/3)(L - r\,\delta_1)^3\big]\mathrm{d}u_3/\mathrm{d}t \\ &-(\rho/3)(L - r\,\delta_1)^3\mathrm{d}u_4/\mathrm{d}t \\ &\quad = -r^2u_1u_3\rho(1 - \cos(\delta_1)) - r\rho(L - r\,\delta_1)^2(u_4 - u_3)u_4 \\ &-r\rho u_3(L - r\,\delta_1)[u_2\cos(\delta_1) + u_1\sin(\delta_1)] \\ &-0.5\rho u_3(L - r\,\delta_1)^2[u_1\cos(\delta_1) - u_2\sin(\delta_1)] \end{aligned} \tag{16}$$

$$
\begin{aligned}
&0.5\rho(L - r\ \delta_1)^2[\mathrm{d}u_2/\mathrm{d}t\ \ \cos(\delta_1) + \mathrm{d}u_1/\mathrm{d}t\ \ \sin(\delta_1)] \\
&- (\rho/3)(L - r\ \delta_1)^3[\mathrm{d}u_4/\mathrm{d}t - \mathrm{d}u_3/\mathrm{d}t] \\
&= -0.5\rho u_3(L - r\ \delta_1)^2[u_1\cos(\delta_1) - u_2\sin(\delta_1)] \\
&- 0.5r\rho(L - r\ \delta_1)^2\left(u_4^2 - u_3^2\right) + \mu R^2\alpha
\end{aligned}
\tag{17}
$$

Note that these equations of motion (14–17) are functions of the generalized speeds, their first derivatives, and only one generalized coordinate δ_1. These are solved simultaneously with the following four Eqs. (18–21) to obtain the trajectory of the system in terms of the generalized coordinates of the system.

$$
\mathrm{d}x/\mathrm{d}t = u_1\cos(\psi) - u_2\sin(\psi) \tag{18}
$$

$$
\mathrm{d}y/\mathrm{d}t = u_2\cos(\psi) + u_1\sin(\psi) \tag{19}
$$

$$
\mathrm{d}\psi/\mathrm{d}t = u_3 \tag{20}
$$

$$
\mathrm{d}\delta_1/\mathrm{d}t = u_4 \tag{21}
$$

The example considered by Seffen [10] restricts the translation of the disk. The set of equations can be reduced by setting $u_1 = u_2 = 0$ and $\mathrm{d}u_1/\mathrm{d}t = \mathrm{d}u_2/\mathrm{d}t = 0$ to verify these against the equations presented by Seffen [10].

This approach of deriving dynamics has been extended for a disk with two antennas and a cube with one or two antennas [8]. Use of symbolic manipulation software is highly recommended to obtain and solve these equations.

References

1. Banerjee AK (2016) Flexible multibody dynamics: efficient formulations and applications. John Wiley & Sons
2. Dewalque F, Collette JP, Brüls O (2016) Mechanical behaviour of tape springs used in the deployment of reflectors around a solar panel. Acta Astronaut 123:271–282
3. Guinot F, Bourgeois S, Cochelin B, Blanchard L (2012) A planar rod model with flexible thin-walled cross-sections. Application to the folding of tape springs. Int J Solids Struct 49(1):73–86
4. Kothari H, Lohiya P, Zubair A, Singh S (2020) Satellite structure of Advitiy (second student satellite of IIT Bombay). In: Advances in small satellite technologies. Springer, Singapore, pp 535–544
5. Modi VJ (1974) Attitude dynamics of satellites with flexible appendages-a brief review. J Spacecr Rocket 11(11):743–751
6. Patki A, Jagdale K, Munjal M (2020) Design approach to antenna deployment system for nano-satellite applications. In: Advances in small satellite technologies. Springer, Singapore, pp 523–533

7. Picault E, Marone-Hitz P, Bourgeois S, Cochelin B, Guinot F (2014) A planar rod model with flexible cross-section for the folding and the dynamic deployment of tape springs: improvements and comparisons with experiments. Int J Solids Struct 51(18):3226–3238
8. Ranade AR (2020) Dynamics modeling of deployment of tape spring antennas and its effects on CubeSat dynamics. Master's Thesis, Indian Institute of Technology Bombay
9. Seffen KA, Pellegrino S (1997) Deployment of a rigid panel by tape-springs
10. Seffen KA, Pellegrino S (1999) Deployment dynamics of tape springs. Proc Royal Soc London Ser A Math Phys Eng Sci 455(1983):1003–1048
11. Walker SJ, Aglietti G (2004) Study of the dynamics of three dimensional tape spring folds. AIAA J 42(4):850–856

Analysis and Experimental Validation of Solder Joints of CCGA Packages

B. K. Chandrashekar, Santosh Joteppa, and Vinod Chippalkatti

Abstract Continuous development in electronics industries results in the introduction of many complex electronics packages. Nowadays, digital and frequency modulation technology occupies a major part in commercial, telecommunication, space, and defence electronics applications. Space industries use high-reliable electronics packages like the Ceramic Column Grid Array package (CCGA) which is programmable as per application in satellite modules. These packages have more number of columns for electronics as well as for mechanical strength. As the number of columns increases, mechanical strength increases, but same time screening of columns that are very closely constructed is very difficult. When these packages undergo screening tests like thermal cycling, there may be chances of solder cracks and bending of columns due to temperature variation. Thereafter, the high-level vibration test makes the solder cracks grow faster and bent columns to fail. These grid array packages shall be integrated onto PCB through reflow technology; reflow creates the solder fillets at the bottom portion of columns which mounts upon PCB. During these reflow processes, in any chance the solder fillet is not properly formed, and vibration test followed by thermal cycling may lead to damage of columns in-turn whole package. As the cost of one package itself is very high, damage to multiple packages may result in huge losses to industries. In this paper, two numbers of 1752 columns CCGA (VERTEX 5 CN1752 CCGA) packages are considered which are mounted on 2.6-mm thick, 12-layer polymide PCB. The board along with CCGA packages is placed inside an aluminium housing, and this assembly is considered to perform dynamic analysis like eigen value, sine response and random response analysis and is subjected to vibration tests to study the solder joint behaviour of CCGA columns; observations are discussed in detail.

B. K. Chandrashekar (✉) · S. Joteppa · V. Chippalkatti
Centum Electronics Limited, Bangalore, Karnataka 560106, India
e-mail: chandrashekarbk@centumelectronics.com

S. Joteppa
e-mail: santoshj@centumelectronics.com

V. Chippalkatti
e-mail: vinod@centumelectronics.com

© The Author(s), under exclusive license to Springer Nature Singapore Pte Ltd. 2023

R.S. Priyadarshini and T. Sundararajan (eds.), *Advances in Small Satellite Technologies*, Lecture Notes in Mechanical Engineering, https://doi.org/10.1007/978-981-19-7474-8_16

Keywords Solder joints · CCGA · Copper ribbons · Solder crack · SMD · FEA · Vibration

1 Introduction

The formation of solder joint during attachment of any electronic components is governed by two processes, the wetting and spreading. Wetting process concerns to the formation of a metallurgical bond between the solder and substrate. Spreading describes the voluntary flow of molten solder over a substrate surface. Wetting and spreading processes together form the solderability of the solder-substrate-flux system [1] (Fig. 1).

There are two common types of solder joint technologies used in highly credible electronics applications, through-hole technology and surface mount technology. Surface mount technology (SMT) is relatively a novel one used to attach electronic components to a printed circuit board. SMT provides various advantages over conventional through-hole technology [2]. Eutectic 63Sn 37Pb is used highly as an interconnection material in electronic packaging. It provides a mechanical as well as electrical connection between component and board. Based on the application, the combination like 63Sn-37Pb and 80Sn-20Pb, etc. are also considered. The solder fillet furnishes both electrical continuity and mechanical fastening for surface mount interconnects (Fig. 2).

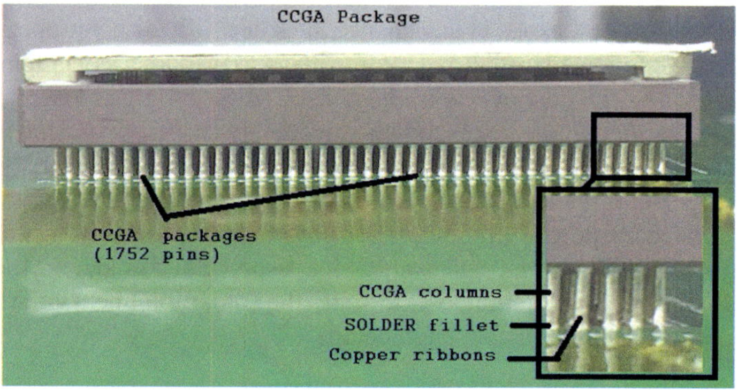

Fig. 1 CCGA columns

Fig. 2 Surface mounting packages

2 Problem Description

During the reflow process, flaws like contaminated surfaces, inadequate flux strength and low processing temperature will result in poor solderability which is often indicated by incomplete joints (cracks) and void formation in the solder fillet, which can degrade the load-bearing capacity as well as electrical functionality [3]. When these are subjected to high-level vibration test followed by thermal cycling, the solder cracks grow rapidly and lead to failure almost immediately. Hence, it is important to study solder joint behaviour for the actual loads through finite element analysis and testing before the actual packages are subjected to screening processes.

3 Finite Element Analysis

Finite element method is a numerical technique used extensively to solve complex engineering problems. In this work, the following analyses are performed to analyse the behaviour of the solder joints,

(1) Eigen value analysis (Modal analysis) to evaluate the fundamental frequencies and modes.
(2) Sine response analysis to check the maximum amplification and vibration signature.
(3) Random response analysis to check the dynamic fluctuation over a confined frequency bandwidth [4].

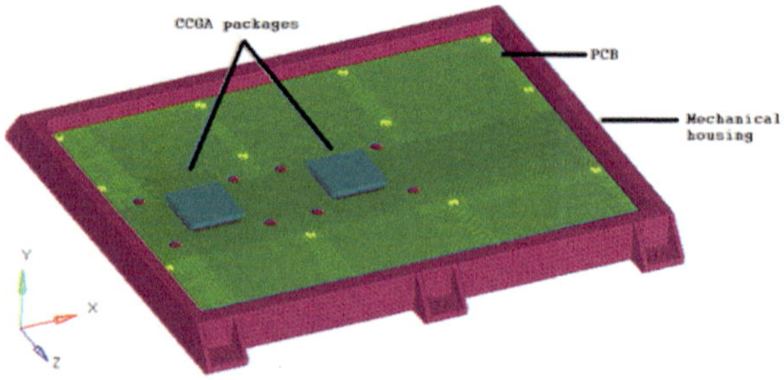

Fig. 3 Finite element model of the assembly

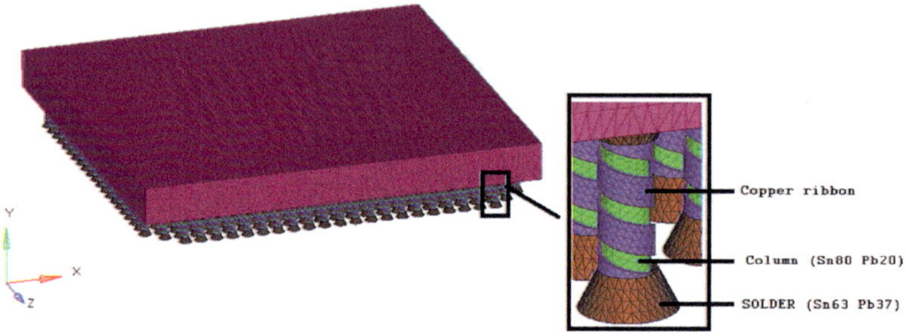

Fig. 4 Finite element model of CCGA columns

3.1 Finite Element Model

The PCB with two numbers of 1752 columns CCGA packages is modelled using two-dimensional shell elements and three-dimensional tetrahedron elements. The PCB along with CCGA packages is placed inside the aluminium housing (refer to Fig. 3) since PCB cannot be mounted directly on armature of the vibration machine. The grid analysis has been performed to optimize the FE model, and standard values of quality parameters like Warpage, Aspect ratio, Tet-collapse and Jacobian are maintained to achieve accurate results [5] (Figs. 4, 5 and Table 1).

3.2 Load Specifications

3.3 Eigen Value Analysis Results

The first fundamental frequency for the assembly is observed at 575 Hz, which is on PCB away from the CCGA packages. Similarly, mode 6 and mode 7 are observed on CCGA packages along with PCB at 838 Hz and 895 Hz, respectively (Fig. 6).

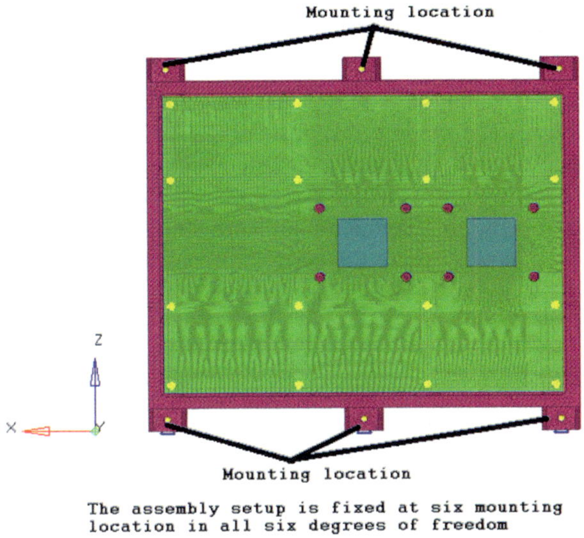

Fig. 5 Boundary condition and excitation location for random response analysis

Table 1 Element type details

Serial number	Element type	Parts	Number of elements
1	2-dimensional shell element	PCB	471,064
2	3-dimensional tetrahedron element	CCGA packages and Housing	11,262,069
3	1-dimensional BEAM element	M3 bolts	16
4	RBE2 elements	Coupling elements	38

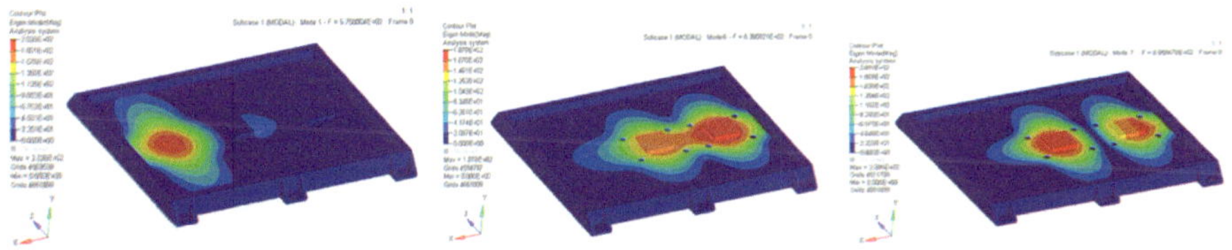

Fig. 6 Eigen value analysis result plot (from left to right—Mode 1, Mode 6 and Mode 7)

3.4 Sine Response Analysis Results

Since the first fundamental frequency of the assembly is observed at 575 Hz, the responses will be flat for sine load as the sine response load is in a frequency bandwidth between 5 and 100 Hz. Hence, sine response analysis is not performed.

3.5 Random Response Analysis Results

The random response analysis is performed with a base excitation for the load as specified in Table 2. The random response 132.31 g^2/Hz (85.5 grms) is observed on the CCGA package which is mounted at the centre of PCB. Since the observed acceleration level is high, it is decided to apply adhesive at all four corners of the packages for safe operation. The maximum RMS stress of 3.94 MPa is observed on the column at the junction where the copper ribbon is attached to the column and solder fillet (Fig. 7).

Table 2 Vibration test specification

Resonance search test	
Frequency (Hz)	Amplitude
5–2000	0.5 g
Sweep rate	2 oct/Min
Sine response specification	
Frequency (Hz)	Amplitude
5–22	12.4 mm (peak-peak)
22–70	25 g
70–100	20 g
Sweep rate	2 oct/min
Random response specification	
Frequency (Hz)	Normal to mounting plane
20–100	+3 dB/oct
100–1000	0.30 PSD (g2/Hz)
1000–2000	−6 dB/oct
Overall	20.8 Grms
Duration	2 min/axis

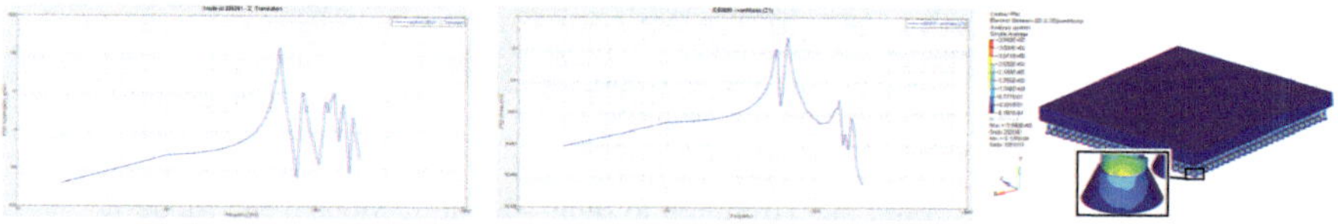

Fig. 7 Random response results plots (from left to right—acceleration plot, RMS stress plot, RMS stress contour)

4 Dynamic Test

During the thermal cycling test, the CCGA columns and solder fillets experience temperature variation, which weakens the pre-existing voids and cracks. Thus, post thermal cycling the assembly shall be subjected to high-level vibration test to identify the failures [6, 7].

The below vibration tests are performed to study the dynamic behaviour of the CCGA columns and solder joints,

(1) Low-level sine (LLS)
(2) Sine response test
(3) Random response test.

The LLS vibration test is performed by applying 0.5 g load for a frequency bandwidth from 5 to 2000 Hz to identify the resonance frequencies (natural frequencies) in the assembly.

The sine and random response tests are performed as specified in the below sequence,

- Pre LLS test—Sine response test—Post LLS test
- Pre LLS test—Random response test—Post LLS test.

Then, the board shall be subjected to detailed visual inspection under the microscope to find any physical changes in columns and solder fillets. The analysis results are referred for visual inspection to reduce the time consumption.

4.1 Test Set up

Vibration test is performed as per the specification shown in Table 2 in all three directions (X, Y & Z) to study the dynamic behaviour of PCB and its effect on CCGA packages. Five number of accelerometers are used to extract the response (two accelerometers on each CCGA packages and three accelerometers on PCB near to CCGA packages). A qualified circular fixture made of aluminium [8, 9] (475 mm in diameter and 40 mm thick) is employed to fix the assembly to the vibration shaker (Figs. 8 and 9).

In this paper,

- the test results when the assembly is loaded along normal to mounting plan (Y direction) are discussed since it is the critical loading direction for PCB and CCGA packages;
- from sine and random response tests, peak response on the CCGA package is discussed; and
- the pre and post-LLS vibration test values for sine and random response tests are tabulated in Table 3 (Figs. 10, 11 and 12).

Fig. 8 Vibration shaker

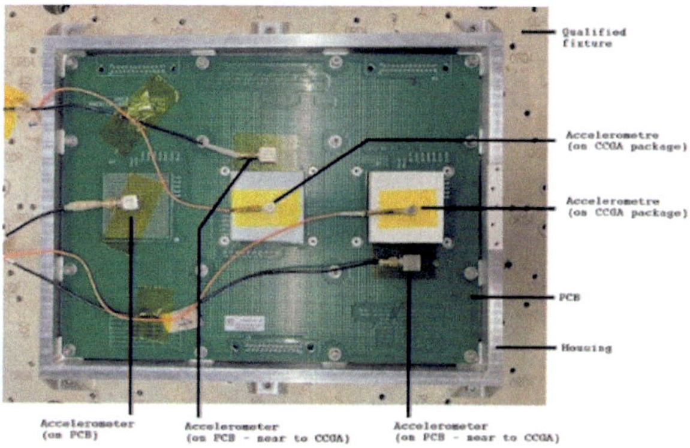

Fig. 9 Vibration test setup

Table 3 Analysis and test result details

Sl. No.		Description	Fundamental frequency (Hz)		Acceleration (g)		RMS stress (MPa)
1	FE analysis	Eigen value analysis	575		–		–
		Random response analysis	575		85.5 grms		3.94
2	Dynamic test	LLS	619		–		
		Sine response test	*Pre-Sine*	*Post-Sine*	*Pre-Sine*	*Post-Sine*	–
			590	590	12.54	12.42	
		Random response test	*Pre-Random*	*Post-Random*	63.12 grms		–
			590	588			
3	Visual inspection		No cracks on solder fillets and column bend				

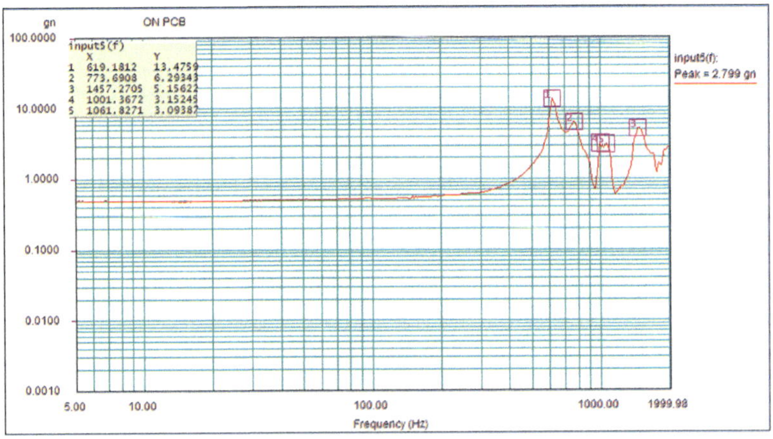

Fig. 10 Low-level sine (LLS) response plots

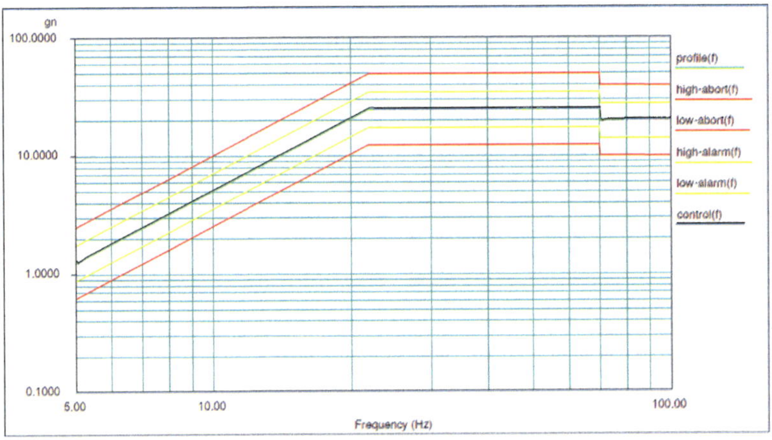

Fig. 11 Sine response test plots

4.2 *Visual Inspection*

Detailed visual inspection is performed under the microscope to identify any major changes or damage in the CCGA columns and solder fillet. It is observed that there is no column bend, column cracks, solder cracks and solder detachments. Later the CCGA packages are subjected to electrical tests, and it is found that both packages are working fine [10–14] (Fig. 13).

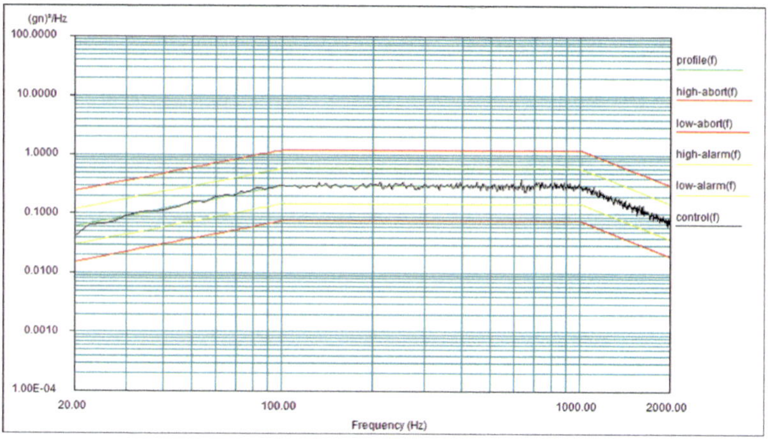

Fig. 12 Random response test plots

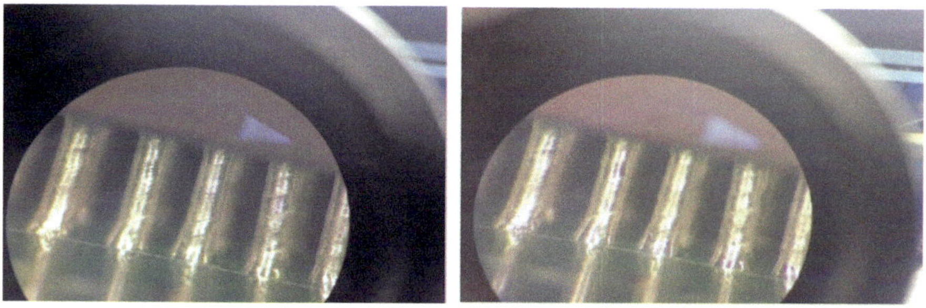

Fig. 13 Visual inspection of CCGA columns and solder fillet under 10 × microscope after vibration tests (Left—before vibration test; Right—after vibration test)

5 Conclusion

The RMS stresses obtained from the FE analysis are well within the yield limit of the material used to design columns and solder (yield strength is 26 MPa). The % error between pre- and post-vibration tests are also well within the allowable limits (10%), but the acceleration amplification is quite high; hence corner stacking is proposed for future studies to avoid amplification. The visual inspection after each vibration test depicts no observation such as crack initiation, cracks growth and solder detachments. Table 3 shows the analysis and testing results in detail.

References

1. Vianco PT (1999) Solder mounting technologies for electronic packages. ResearchGate, Affiliation: Material Joining Dept., Sandia National Laboratories
2. Vianco PT (2017) Understanding the reliability of solder joints used in advanced structural and electronics applications: part 1—filler metal properties and the soldering process. Sandia National Laboratories Albuquerque, NM
3. Goddard Space Flight Center, February 22, 2011. Ceramic Column Grid Array Design and Manufacturing Rules for Flight Hardware, GSFC-STD-6001
4. Steinberg DS (1999) Vibration analysis for electronics, 3rd edn. A Wiley-Interscience publications
5. Gokhale NS (2008) Practical finite element analysis. Published by Finite to Infinite
6. Ghaffarian R (2005) Reliability of CGA/LGA/HDI package board/assembly (revision A). Jet Propulsion Laboratory Pasadena, California
7. Ghaffarian R (2008) Thermal cycle reliability and failure mechanisms of CCGA and PBGA assemblies with and without corner staking. IEEE Trans Compon Packag Technol 31(2)
8. Joteppa S, Chippalkatti VS (2012) Design and qualification of hybrid micro circuit packages for higher vibration loads for aerospace and defence application. Int J Aerosp Sci 2(3):106–123, ISSN: 2169-8872; e-ISSN: 2169-8899
9. Chandrashekar BK, Joteppa S, Bakale KR, Chippalkatti VS (2017) Life estimation of HMC package under dynamic load condition and correlation of analysis and test results. J Mech Eng Autom 7(5):172–178, p-ISSN: 2163-2405; e-ISSN: 2163-2413
10. Fleisher J, Willing W (2011) Study of column grid array components for space systems. Northrop Grumman Corporation
11. Irvine MH (2014) The international voice of the test, assembly, and packaging community. SEMICON West 5(7)
12. Park T-Y, Jeon S-H, Kim S-J, Jung S-H, Oh H-U (2018) Experimental validation of fatigue life of CCGA 624 package with initial contact pressure of thermal gap pads under random vibration excitation. Hindawi Int J Aerosp Eng 2018:2697516
13. Delta technical data sheet, Sn63/Pb37, 215D, Rev. C, 11/15
14. Xilinx; Virtex-5QV FPGA Packaging and Pinout Specification, UG520 (v1.7) October 11, 201

Challenges in the Dynamic Condensation and Internal Response Prediction of Small Satellites

Narendra Nath, K. P. Venkateswaran, T. Sundararajan, and S. Geetha

Abstract Emerging small satellite market faces many challenges mainly to protect their payload during the launch. Satellite team has to share their finite element model to the launch vehicle team for the various analysis like launch vehicle dynamic characteristics, Coupled Load Analysis (CLA), dynamic envelop studies, etc. FE model contains sensitive details about the satellite, and dynamic condensation of the satellite is the only way to protect satellite details. This also helps in reducing the number of degrees of freedom of the satellite finite element model. Craig-Bampton technique is the most popular method to reduce the satellite into a mass and stiffness matrix with the required number of modes. Satellite team needs a physical node at critical locations to study the responses, and these nodes need to be retained while reduction. This paper examines the Craig-Bampton reduction method and discusses how to compute the response at various locations of the satellite. A frequency response analysis with reduced model and full model is also compared in this paper.

Keywords Craig-Bampton · Modal condensation · Frequency response · Small satellite

1 Introduction

Small satellite manufacturing is the fastest growing market with a market share of 2.8 billion dollars by the year 2020 and would be 20% substantial growth from 2020 to 2025 [1]. As the market share increases, the challenges also ballooned to protect their design and internal payload details from other competitive companies. Satellite manufacturer has to share their finite element model (FEM) to the launch vehicle team for the overall launch vehicle dynamic analysis as well as transient analysis for predicting response at the critical location, from where the satellite ground test vibration levels used to be generated. Various condensation techniques are currently being used by various agencies to reduce the finite element model to mass and

N. Nath (✉) · K. P. Venkateswaran · T. Sundararajan · S. Geetha
Vikram Sarabhai Space Centre, ISRO, Thiruvananthapuram 695022, India
e-mail: narendra2232in@gmail.com

© The Author(s), under exclusive license to Springer Nature Singapore Pte Ltd. 2023

R.S. Priyadarshini and T. Sundararajan (eds.), *Advances in Small Satellite Technologies*, Lecture Notes in Mechanical Engineering, https://doi.org/10.1007/978-981-19-7474-8_17

stiffness matrices, and among all, Craig-Bampton technique is most popular and widely recognized [2].

Craig-Bampton mathematical formulation is extensively used in aerospace applications, where large finite element models cannot be directly integrated or increases the size of the whole model, which ultimately takes a large time to solve [3]. The mathematical formulation of Craig-Bampton is often used for:

(1) Reducing the number of degrees of freedom of the actual model and supply to other agencies for coupled load analysis (CLA) and dynamic characterization of the launch vehicle [4].
(2) Modal synthesis, where the satellite model can be attached to interfacing structure like payload adapter and multi-satellites interface adapter.

Small satellite payloads need to be vibration tested as a qualification plan, and it becomes mandatory to find the response levels at the critical package locations. These responses can be estimated through CLA if the satellite is reduced with a single interface node, and the responses of internal nodes of interest are estimated using various ways. Another way is to condense the satellite model with the degrees of freedom of the desired nodes (internal points) where the responses are of interest. Detailed condensation technique and efficacy of this technique is brought out in this paper. Frequency response will also be compared between the full and condensed model for one of the critical payload locations. Various techniques will also be discussed to calculate internal point response of the satellites.

2 Mathematical Formulation of the Craig-Bampton Technique

A brief description of the Craig-Bampton theory will be discussed in this section [5].

The basic equation of motion is generally considered below ignoring structural damping

$$[M_{AA}]\{\ddot{u}_A\} + [K_{AA}]\{u\} = \{F(t)\} \tag{1}$$

The Craig-Bampton transformation is defined as:

$$\{u_A\} = \left\{\frac{u_b}{u_L}\right\} = \begin{bmatrix} I & 0 \\ \emptyset_R & \emptyset_L \end{bmatrix}\left\{\frac{u_b}{q}\right\} \tag{2}$$

where, u_b = boundary degrees of freedom (DOF), u_L = internal (leftover) DOF's, ϕ_R = rigid body vector, ϕ_L = fixed base mode shapes and q = modal DOF's, $\begin{bmatrix} I & 0 \\ \phi_R & \phi_L \end{bmatrix}$ = ϕ_{cb} = C-B transformation matrix.

Combining Eqs. (1) and (2) and pre-multiplying by $[\phi_{cb}]^T$

$$\emptyset_{cb}^{T}[M_{AA}]\phi_{cb}\left\{\frac{\ddot{u}_b}{\ddot{q}}\right\}+\emptyset_{cb}^{T}[K_{AA}]\phi_{cb}\left\{\frac{u_b}{q}\right\}=\emptyset_{cb}^{T}\left\{\frac{F_b}{F_L}\right\} \tag{3}$$

Defining the C-B mass and stiffness matrices as

$$[M_{cb}]=\emptyset_{cb}^{T}[M_{AA}]\phi_{cb}=\begin{bmatrix}\mathrm{M}_{bb} & M_{bq}\\ M_{qb} & \mathrm{M}_{qq}\end{bmatrix} \tag{4}$$

$$[K_{cb}]=\emptyset_{cb}^{T}[K_{AA}]\phi_{cb}=\begin{bmatrix}\mathrm{K}_{bb} & 0\\ 0 & K_{qq}\end{bmatrix} \tag{5}$$

Re-write Eq. (3) using Eqs. (4) and (5)

$$\begin{bmatrix}\mathrm{M}_{bb} & M_{bq}\\ M_{qb} & \mathrm{M}_{qq}\end{bmatrix}\left\{\frac{\ddot{u}_b}{\ddot{q}}\right\}+\begin{bmatrix}\mathrm{K}_{bb} & 0\\ 0 & K_{qq}\end{bmatrix}\left\{\frac{u_b}{q}\right\}=\left\{\frac{F_b}{0}\right\} \tag{6}$$

When structural damping is considered, dynamic equation motion can be written as

$$\begin{bmatrix}M_{bb} & M_{bq}\\ M_{qb} & I\end{bmatrix}\left\{\begin{matrix}\ddot{u}_b\\ \ddot{q}\end{matrix}\right\}+\begin{bmatrix}0 & 0\\ 0 & 2\zeta\omega\end{bmatrix}\left\{\begin{matrix}\dot{u}_b\\ \dot{q}\end{matrix}\right\}+\begin{bmatrix}K_{bb} & 0\\ 0 & \omega^2\end{bmatrix}\left\{\begin{matrix}u_b\\ q\end{matrix}\right\}=\left\{\begin{matrix}F_b\\ 0\end{matrix}\right\} \tag{7}$$

where $2\zeta\omega =$ modal damping.

3 Analysis

A small satellite model with the dimension 378 mm × 378 mm × 453 mm with 32.5 kg of mass was considered for the study. All the electronic packages were modelled as lumped mass at the centre of gravity and connected using rigid beams. Nodes on the interface ring with the launch vehicle were brought to a single node. While condensation of the model, this single node was treated as an interface node. This satellite model was reduced to a single interface node using the Craig-Bampton technique through NASTRAN solver for 60 modes [6]. Actual/full satellite model and Craig-Bompton reduced models are analysed for the base fixed conditions. Frequency, mass and inertial parameters are compared to check the effectivity of the technique.

Frequency response analysis also was carried out to understand the effect of reduction. Full model, as well as reduced models, were excited in X-axis (see the axis definition in Fig. 1) and responses at critical package locations were compared.

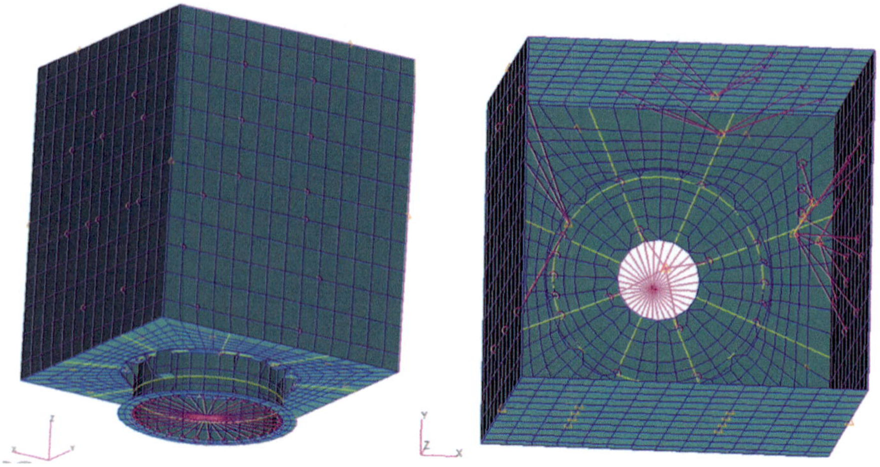

Fig. 1 Small satellite FE model is shown in left figure and right figure shows satellite internal details with mass lumping locations connections

4 Results and Discussions

Craig-Bompton technique was used to reduce the given model into a single node. Comparison of frequencies between the full and reduced model for the base fixed boundary condition for first 60 modes (1438 Hz) are plotted in Fig. 2. Figure 2 illustrates that both the full and reduced models are in very good agreement. Since the first base fixed frequency is 82 Hz, the curve starts from the 82 Hz. Mass and moment of inertia of reduced and full satellite model are very well comparable, which show reduction in technique efficiency.

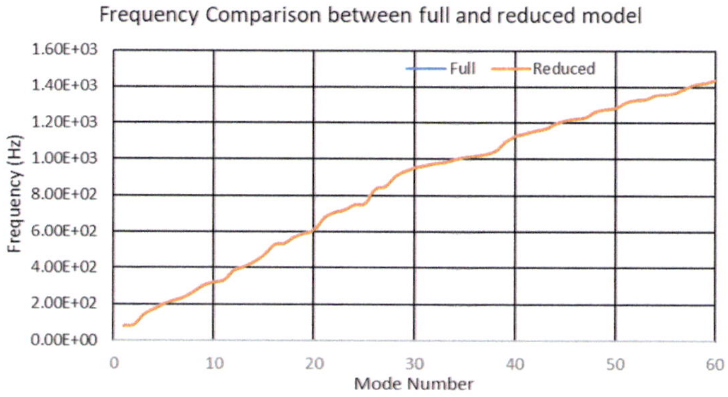

Fig. 2 Shows frequency comparison between the full and reduced model for the base fixed condition of the satellite

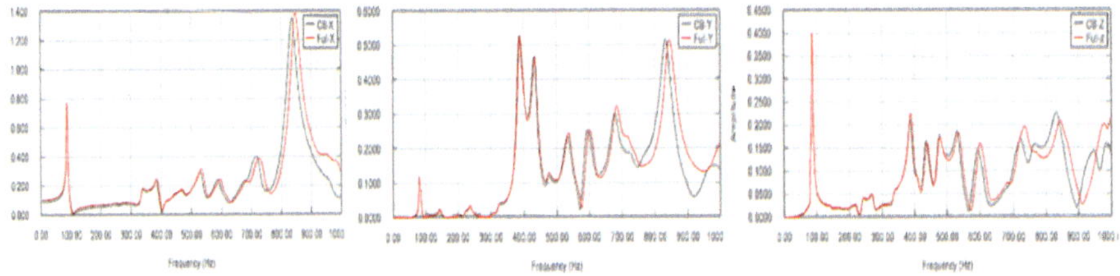

Fig. 3 Frequency response comparison between the full and reduced model for the excitation in X-axis, at **a** node 151,672 in X-axis, **b** node 151,672 in Y-axis and **c** node 151,672 in Z-axis

Figure 3 shows frequency response comparison in X, Y and Z directions at node 151,672 which is one of the critical package location nodes, between the full model and reduced model with internal nodes retained; a clear difference in response between full and reduced models is seen from 500 Hz frequency onwards. The peak amplitude 1.4 is seen in X-directions; however, Y and Z response amplitudes are 0.4 and 0.6, respectively. Higher frequency mismatch is due to not including the residual vectors, hence it is recommended to include the residual vector while reducing as well as during transient analysis.

5 Internal Point Response Computation

Satellite model contains various sensitive packages like camera or sun sensors where responses need to be computed to define test levels. This is another challenging area which needs to be taken care while computing the internal point responses. This section is going to introduce various techniques which can be incorporated for computing the internal point response.

The physical node extraction technique is a very popular technique in which critical locations nodal degrees of freedom are kept free while doing generalized dynamic reduction or component mode synthesis. These physical nodes in the reduced model are to be used for the extraction of response directly while doing coupled load analysis with the launch vehicle. Another method is to make restitution matrix which is the mode shape of the internal points. Response at the launch vehicle satellite interface is calculated through CLA, and this response is extrapolated to the internal point using this restitution matrix. The technique is very effective when the number of internal points are quite high, especially large satellites.

In some of the cases where the satellite is supplied to a single node without having given the extraction node or restitution points, it is difficult for the launch vehicle agencies to extract the internal point response. In such a case, launch vehicle agencies can only supply the satellite base response to the satellite team and satellite team has to apply this given base response as transient to satellite base and extract response to the desired locations. Certain cases where satellite reduced FE model is not at supplied, a lumped mass technique is used in which satellite mass is lumped at the centre of gravity of the satellite and responses are extracted.

6 Conclusions

Satellite reduction technique was discussed in this paper, and a detailed formulation of the most popular Craig-Bampton reduction technique was demonstrated. The full FE model and reduced model is compared for the frequency and found that they are in excellent agreement to each other. Similarly, frequency response analysis was also brought out and mismatch in frequency response is attributed to residual vector which needs to be included while analysing the model. Another challenge is how to compute the internal point response at the critical/sensitive package location. A detailed internal point response computation technique is also brought out in this paper.

Acknowledgements We would like to thank Director, VSSC, for giving me the opportunity to publish work. We are sincerely thankful to the editorial board of VSSC to provide the necessary tools and direction. Last but not least, we convey my sincere gratitude to the internal reviewer for their frank and honest review to improve the quality of work.

References

1. Small satellite market by mass (nanosatellite, microsatellite, minisatellite, CubeSat), subsystem (satellite bus, payload, satellite antenna, solar panels), application, NGEO orbit, end user, frequency, and region—global forecast to 2025; https://www.marketsandmarkets.com/Market-Reports/small-satellite-market-150947396.html?gclid=CjwKCAjwnef6BRAgEiwAgv8mQQCHMSvTJMTQMzu7arBB40iERoaNCxD1oZ9HYYKVPsatMjnAoqmz-xoCjxYQAvD_BwE
2. Pray C, Blelloch P, Thomas G, McNelis M, Saurez V, Otten K (2008) Structural dynamic analysis of a spacecraft multi-DOF shaker table. In: 25th space simulation conference 2008, (October 2008), pp 281–298
3. Young JT (2000) An introduction to boudary node functions, base shake analysis, load transformation matrices, modal synthesis and much more. https://femci.gsfc.nasa.gov/craig_bampton/Primer_on_the_Craig-Bampton_Method.pdfAuthor
4. Craig RR, Bampton MCC (1968) Coupling of substructures for dynamic analysis, HAL archives-ouvertes. https://hal.archives-ouvertes.fr/hal-01537654
5. MD NASTRAN (2010) Dynamic analysis user's guide, 2010. MSC Software Corporation
6. MD NASTRAN (2010) Quick reference guide, 2010. MSC Software Corporation

Design and Additive Manufacturing of a Mechanical Chassis for Small Satellite Launch Vehicle Inertial Navigation Package

Tony M. Shaju, D. Syamdas, G. Nagamalleswara Rao, K. Pradeep, and Joji J. Chaman

Abstract The demand for a small satellite launch vehicle for ISRO has been ever increasing due to growing launch requirements from the rising number of industrial and academic players who are venturing into design and fabrication of smaller satellites for their specific applications. Therefore, a new lightweight navigation system needs to be designed for catering to the requirements of a low weight small satellite launch vehicle. The major mechanical component of a navigation package is the cluster (chassis) which is used for mounting gyroscopes and accelerometers required for providing information about the position and attitude of the launch vehicle. The cluster adds a good percentage to the total mass of the package. As a first step towards proving the technical feasibility of AM route, the CNC machinable cluster was 3D printed along with test coupons, and their structural and thermal characteristics were studied. Then, improvements were made on the cluster to increase one of the normal mode frequencies using design for additive manufacturing philosophy. This modification also helped to reduce the mass, which was made possible with the flexibilities offered by additive manufacturing in terms of manufacturing constraints. The modal analysis carried out on the modified cluster proved the improvement in performance of the cluster. The modified cluster was successfully 3D printed, followed by testing to meet all geometric and performance requirements.

Keywords Additive manufacturing · Chassis design · Design for additive manufacturing · Direct metal laser sintering · 3D printing

1 Introduction

The growth of the additive manufacturing (AM) process over the past few years as a mainstream manufacturing process has been tremendous. Compared to conventional methods such as machining and injection moulding, AM offers unparalleled potential in manufacturing of complex shapes and customized geometry. The major advantages

T. M. Shaju (✉) · D. Syamdas · G. Nagamalleswara Rao · K. Pradeep · J. J. Chaman
ISRO Inertial Systems Unit, Trivandrum, India
e-mail: tonymshaju@gmail.com

© The Author(s), under exclusive license to Springer Nature Singapore Pte Ltd. 2023

R.S. Priyadarshini and T. Sundararajan (eds.), *Advances in Small Satellite Technologies*, Lecture Notes in Mechanical Engineering, https://doi.org/10.1007/978-981-19-7474-8_18

of AM include part consolidation, weight reduction, functional enhancement, part customization, etc. [1]. AM particularly supports low volume manufacturing [2], as manufacturing these components via traditional manufacturing methods might turn out costly. This also creates opportunities for supply chain simplification [3]. AM helps in reducing raw material cost for components where the input stock is machined away to less than 10% [4]. Current advances in automated manufacturing technologies have also enabled hybrid manufacturing (HM) processes that combine the complementary capabilities of AM and subtractive manufacturing [5].

When using conventional design methodology, one has to take into consideration the limitations of the CNC machining processes like geometric complexity, tool access to interior locations, etc. But AM can build different types of intricate features without being limited by the CNC machining design constraints [6]. A design philosophy called Design for Additive Manufacturing methodology has been developed and adopted, and still being improved for the purpose of designing efficient and performance-enhanced structures for AM [7]. While AM provides huge design potentials, the geometric freedom available is not unlimited [8].

Determining the properties of the powder used for AM, as well as the properties of the resulting bulk metal material, is necessary for qualification of the AM component [9]. Powder testing methods include chemical composition check and a range of complex physical measurement techniques like laser diffraction particle sizing and morphological analysis [10]. Testing for mechanical properties like Yield strength, Young's modulus, shear strength, fracture toughness, hardness, etc., helps in adding confidence to the material and process being used for building the component. The use of NDT methods for the substantiation of quality and structural integrity of additively manufactured parts are necessary in this respect. X-ray computed tomography (CT) is the most encouraging method for the inspection of parts with complex geometries [11].

2 Mechanical Design of Cluster

The sensor cluster (or cluster) is used to mount navigation sensors like gyroscopes and accelerometers and their associated electronic components at required locations. The cluster package is then assembled in the cluster housing suspended through a set of vibration elastomeric isolators, which is eventually mounted on the equipment bay of the Small Satellite Launch Vehicle. The design drivers for the cluster are its stiffness and thermal performance to ensure optimum sensor performance during launch environment. The requirement is that inertial sensor errors does not vary in time due to dynamic inputs to the cluster due to vibrational and thermal loading. The rocket motor acoustics induced random vibration input to the cluster is attenuated by suspending the cluster on elastomer-based vibration isolators. The inertial sensors used in launch vehicle navigation generate heat while they are at work. These sensors perform optimally in a given temperature range only. Thus, a thermal control needs to be implemented within the cluster to maintain thermal equilibrium, as external

Table 1 Chemical composition of AlSi10Mg material

Element	Al	Si	Cu	Mn	Mg	Zn	Fe
Weight %	Bal	9–11	≤0.1	0.05	0.45–0.6	0.05	≤0.55

Table 2 Typical mechanical properties of AlSi10Mg material

Property	Horizontal	Vertical
Yield Strength in MPa	248	228
Tensile strength in MPa	386	412
Elongation in %	8.6	7

conditions vary as the launch vehicle ascends. However due to elastomeric suspension, the conduction paths do not exist, and the only means of transporting heat is radiation, which demands more area, which would in turn increase the mass. Therefore, a material with high specific stiffness and stable thermal properties is preferred for manufacturing a cluster. For CNC machining, aluminium alloy 6061 is used.

For the feasibility of using traditional manufacturing methods, the cluster was initially designed keeping in mind the design constraints imposed by 5-axis CNC machining. This design itself is complicated due to the closed structure of the cluster and assembly challenges caused by small available design volume of the cluster. Initially, this design was taken for 3D printing to assess the feasibility of the process and to make sure the process meets all design and functional requirements. AlSi10Mg alloy was used for additive manufacturing of the cluster owing to its good mechanical properties like Yield strength and Young's modulus, high thermal conductivity, and low density [12]. The chemical composition of AlSi10Mg material is given in Table 1. Typical mechanical properties of AlSi10Mg material [13] are given in Table 2. The 3D printing operation was performed successfully.

Later, design modifications (Figs. 1, 2, 3, and 4) were done on the cluster using design for additive manufacturing philosophy for decreasing the mass and increasing the first normal mode frequency of the cluster. The design of the centre stem and mounting locations of the accelerometers underwent major modifications.

Modifications were done on two other locations as well. This helped in further reduction in mass. The modified cluster was 3D printed along with test coupons using the same process and the same machine.

The geometrical tolerances of areas where the sensors are mounted on the cluster need to be very stringent. However, this is difficult to be achieved with AM. Hence, stock material was added at these critical locations to allow post machining of the cluster to achieve high precision geometric tolerances. Build simulation was performed to assess the distortion that forms on the structure during the printing operation.

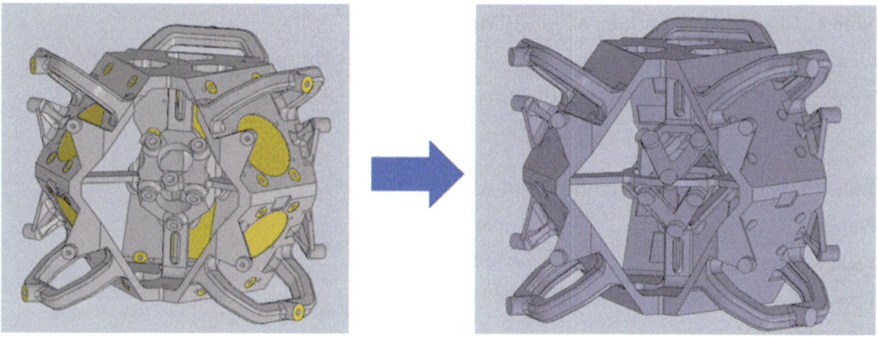

Fig. 1 The CNC machinable design and the modified design of the cluster

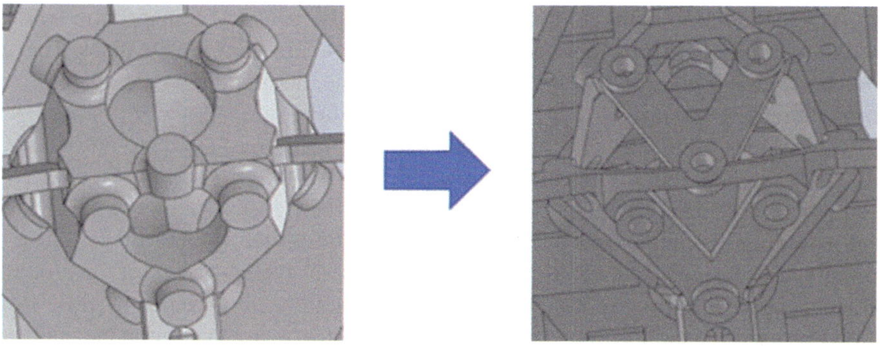

Fig. 2 Modifications done at the centre stem which is the location for mounting accelerometers

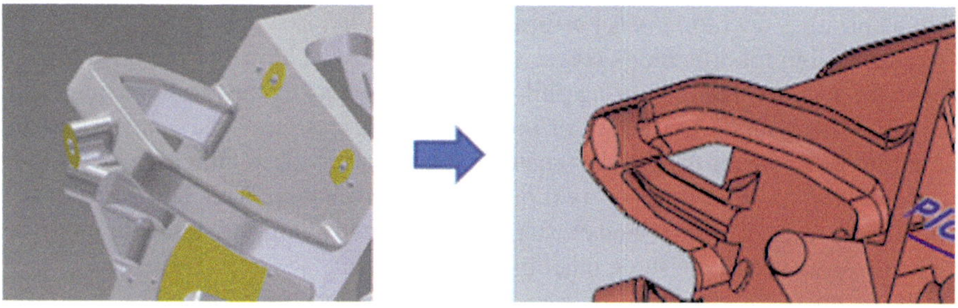

Fig. 3 Conversion of the cross section of the cluster handle to C section

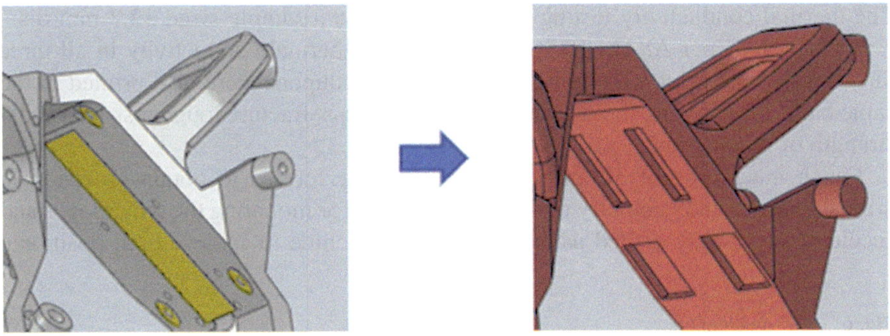

Fig. 4 Modification at the internal faces by addition of pockets which are not machinable by CNC milling

3 Additive Manufacturing and Testing of Cluster and Coupons

The CNC machinable cluster and the modified cluster was 3D printed using DMLS technology on an EOS M280 machine. The build orientation was selected in such a way that the part height was minimized. The build was performed on a preheated platform to reduce the formation of residual stresses in the structure. The build platform with the cluster and the test coupons is shown in Fig. 5

The 3D printed clusters are shown in Figs. 6 and 7. Three each test coupons were printed on the same build in *XY* and *Z* directions in order to use them for mechanical testing. The test coupons were made extra longer to enable a cutting of samples for thermal conductivity tests. The tensile testing was conducted in UTM, and the Young's modulus was analytically calculated from force vs deflection graph. The yield strength and percentage elongation at break also were obtained during the test.

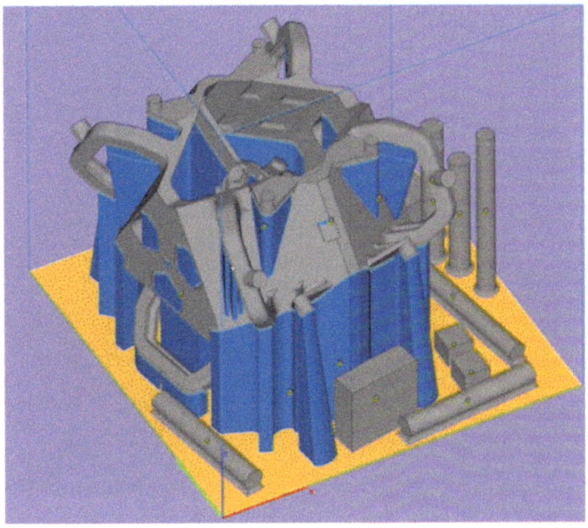

Fig. 5 Build orientation of the cluster along with respective test coupons

The thermal conductivity testing was conducted on a sample sized 12.7 mm dia * 3 mm length as per ASTM E1461 standard. The thermal conductivity in all three directions of the print were measured. Other test coupons were also printed in the same build for enabling the measurement of hardness, fracture toughness, and shear strength of the material.

Finish machining of 3D printed cluster was performed to obtain precision in critical features and geometry of the cluster. Holes for mounting the gyroscopes and accelerometers were drilled using a jig boring machine to achieve high positional

Fig. 6 CNC machinable cluster printed by Direct Metal Laser Sintering

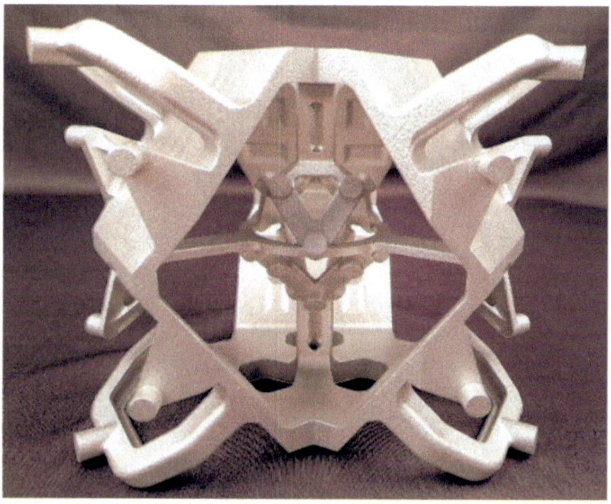

Fig. 7 Modified cluster 3D printed by Direct Metal Laser Sintering

accuracy. CT scan of the 3D-printed cluster was performed to detect formation of pores in the internal structure of the component during the printing process.

4 Results and Discussions

The modification done on the CNC machinable cluster using design for additive manufacturing methodology yielded an increase in the first normal mode frequency and a decrease in the mass of the cluster. A lower bound Young's modulus value of 65GPa was used for AlSi10Mg material during simulation for maintaining a safety margin. The first normal mode frequency of the CNC machinable cluster and the modified 3D printed cluster, which is in the thrust direction of the launch vehicle is shown in Fig. 8. Details of mass reduction and frequency increase are given in Table 3.

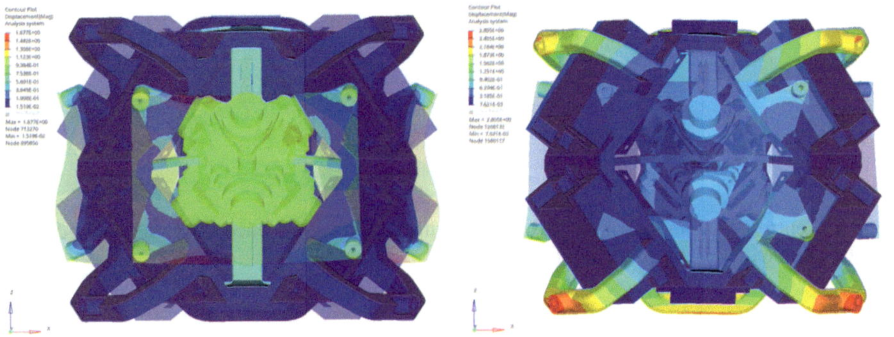

Fig. 8 First normal mode frequency of the CNC machinable cluster and the modified 3D printed cluster in the launch vehicle thrust direction

Table 3 Increase in centre stem thrust mode frequency due Design for Additive Manufacturing-based design change

Design	CNC Machinable cluster	Modified AM cluster
Material	AA6061/6351	AlSi10Mg
Young's Modulus used for simulation	70GPa	65GPa
Centre stem thrust mode frequency	1352 Hz	1427 Hz
Percentage reduction in Centre stem mass	–	40%
Overall Reduction in cluster mass	–	13%

The centre stem of the cluster where maximum amount of mass reduction was done is shown in Fig. 9 This mass reduction was the reason for the substantial increase in the first normal mode frequency of the cluster.

Build simulation performed on the cluster CAD model using SimufactTM software yielded total displacement results for different locations of the cluster. The maximum predicted total displacement (Fig. 10) at any location on the cluster was 1.24 mm.

The CNC machinable cluster and the modified cluster were successfully 3D printed using DMLS technology. The printing process of one cluster took 70 h for

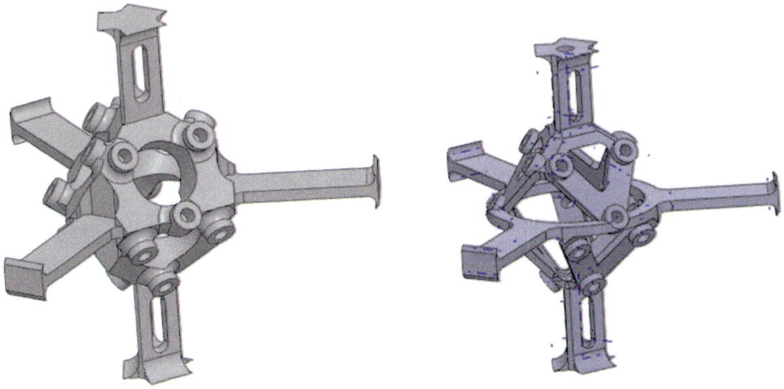

Fig. 9 Centre stem comparison of the CNC machinable and the modified clusters. Mass reduction of 40% was achieved due to the modification

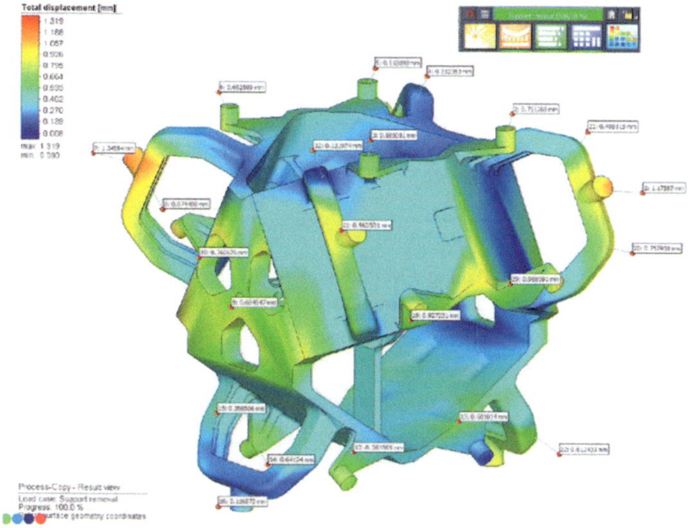

Fig. 10 Simulation on the total displacement of the critical features of the cluster

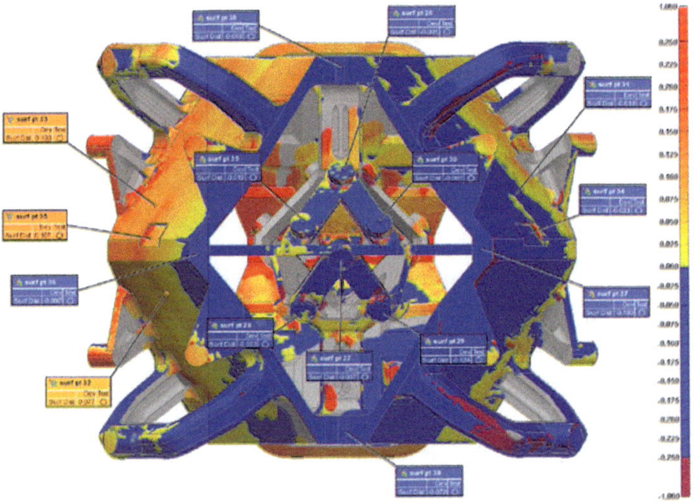

Fig. 11 3D scanning results for the 3D printed modified cluster

completion, against time taken for CNC machining which is about 3 weeks. After printing process, 3D scanning was performed, and the scanned data was compared with original CAD file of the cluster to study the deviations of the printed cluster with respect to the original geometry. A maximum deviation of 260 microns (Fig. 11) was observed. Although measured values of distortion are less than the predicted values, the predicted areas of high distortion matched the areas in the cluster where high distortions were measured.

3 each test coupons printed in *XY* and *Z* direction were tested using a UTM to obtain the Yield strength, Young's modulus and percentage of elongation at break. The nominal values (Tables 4 and 5) obtained are close to values reported in technical literature [13]. Thermal conductivity tests were done on small button sized samples cut out from the respective tensile coupons. The results indicated a range of 165–200 W/mK values for different build directions.

Fracture toughness and shear strength for *XY* and *Z* directions of test samples were tested as per ASTM E399 and ASTM E769, respectively, and the results are shown in Table 6. These values are close to the respective values reported for aluminium alloy 6061, which is used to manufacture the CNC machinable cluster.

Table 4 Tensile properties of test coupons in XY direction

Property	*XY*-1	*XY*-2	*XY*-3
Yield strength in MPa	254	244	245
Elongation in %	9.70	9.80	10.00
Reduction area in %	11.25	16.78	12.56
Modulus in GPa	75	73	76

Table 5 Tensile properties of test coupons in *Z* direction

Property	Z-1	Z-2	Z-3
Yield strength in MPa	247	235	235
Elongation in %	5.20	5.30	5.03
Reduction area in %	6.52	9.72	10.05
Modulus in GPa	76	74	76

Table 6 Fracture toughness and shear strength measurements for AlSi10Mg material

Property	*XY*	*Z*
KQ value in MPa$\sqrt{m}$	31.99	23.49
Shear strength in MPa	215	197

Table 7 Hardness measurement of AlSi10Mg test samples in HBW

Sample identification	1	2	3
H1	102	102	104
H2	102	102	99
H3	102	99	102

Hardness values of AlSi10Mg test samples tested as per ASTM E10 are shown in Table 7. Measurements were taken for three samples at three different locations for each sample.

CT scanning results (Fig. 12) for one layer of the cluster are shown in Fig. 5. Scan of each layer of the cluster that is 100 microns apart was obtained and studied to identify potential crack formation. The results showed no porosities or cracks bigger than 100 microns in the entire volume of the cluster.

Fig. 12 A cross sectional view from the CT scan results

5 Conclusions

The following conclusions were obtained as part of the work done on the cluster used for inertial navigation purpose on small satellite launch vehicles:

- The design change carried out on the CNC machinable cluster was inspired by the Design for Additive Manufacturing methodology. This paved the way for a reduction in mass of the cluster and an increase in the frequency of a critical mode of the cluster.
- The mechanical properties of the AlSi10Mg material used for 3D printed were lab tested and found to be similar to or higher than the AA6061 material. The higher measured Young's modulus value compared to the inputs used for simulation studies will further help in increasing the critical mode frequency of the cluster. Thus, the increase in critical mode frequency is contributed both by the advantages of design for additive manufacturing methodology and by the higher Young's Modulus of the material used for 3D printing.
- The thermal property of the AlSi10Mg material and the 3D printing process methodology used, in terms of thermal conductivity is very similar to the AA6061 material used for CNC machining of cluster, and therefore meets the functional requirement of the cluster design.
- Build simulation conducted before printing operation helped in identifying optimum orientation of the cluster for best possible print results with minimum distortion and residual stresses.
- The DMLS technology produced good results in terms of shear strength, fracture toughness, hardness, and lack of cracks and porosity in the internal volume of the cluster.
- The use of AM route for manufacturing of the cluster resulted in reduction of time for realization of the cluster.

References

1. Doubrovski Z, Verlinden JC, Geraedts JMP (2011) Optimal design for additive manufacturing: opportunities and challenges. In: International design engineering technical conferences and computers and information in engineering conference, vol 54860, pp 635–646
2. Rosen DW (2007) Design for additive manufacturing: a method to explore unexplored regions of the design space. In: 2007 International solid freeform fabrication symposium
3. Pereira T, Kennedy JV, Potgieter J (2019) A comparison of traditional manufacturing vs additive manufacturing, the best method for the job. Proc Manuf 30:11–18
4. Sweeney M, Acreman M, Vettese T, Myatt R, Thompson M (2015) Application and testing of additive manufacturing for mirrors and precision structures. In: Material technologies and applications to optics, structures, components, and sub-systems II, vol 9574. International Society for Optics and Photonics, p 957406
5. Mirzendehdel AM, Behandish M, Nelaturi S (2020) Topology optimization with accessibility constraint for multi-axis machining. Comput Aided Des 122:102825

6. Thompson MK, Moroni G, Vaneker T, Fadel G, Ian Campbell R, Gibson I, Bernard A et al (2016) Design for additive manufacturing: trends, opportunities, considerations, and constraints. CIRP Ann 65(2):737–760
7. Ponche R, Kerbrat O, Mognol P, Hascoët J-Y (2014) A novel methodology of design for additive manufacturing applied to additive laser manufacturing process. Robot Comput Integr Manuf 30(4):389–398
8. Kumke M, Watschke H, Vietor T (2016) A new methodological framework for design for additive manufacturing. Virtual Phys Prototyping 11(1):3–19
9. Cooke AL, Slotwinski JA (2012) Properties of metal powders for additive manufacturing: a review of the state of the art of metal powder property testing. https://doi.org/10.6028/NIST
10. Clayton J (2014) Optimising metal powders for additive manufacturing. Met Powder Rep 69(5):14–17
11. Lu QY, Wong CH (2018) Additive manufacturing process monitoring and control by non-destructive testing techniques: challenges and in-process monitoring. Virtual Phys Prototyping 13(2):39–48
12. Kempen K, Thijs L, Van Humbeeck J, Kruth J-P (2015) Processing AlSi10Mg by selective laser melting: parameter optimisation and material characterisation. Mater Sci Technol 31(8):917–923
13. Casati R, Hamidi Nasab M, Coduri M, Tirelli V, Vedani M (2018) Effects of platform pre-heating and thermal-treatment strategies on properties of AlSi10Mg alloy processed by selective laser melting. Metals 8(11):954

Investigation of Change in Dynamic Unbalance of a Reaction Wheel After Random Vibration Test

D. Syamdas, K. Pradeep, Joji J. Chaman, and K. Anilkumar

Abstract Spacecraft elements like reaction wheels are subjected to random vibration tests. These tests are done to simulate launch loads during ascend phase of the launch, to establish the design margins, and to bring out workmanship-related issues. Random vibration tests typically excite all natural frequencies of the structure within the excitation bandwidth. The excitation bandwidth typically ranges between 20 and 2000 Hz. The excited range of frequencies is significantly wide, so that it is practically impossible to design a structure without any natural frequencies within the range of excited frequencies. Hence, they must be designed in such a way that they must perform satisfactorily during these tests even when they are resonated at their natural frequencies. Flywheels of reaction wheels are subjected to very high vibration loads; these loads are transferred to bearing unit through the assembly joints. The integrity of these joints plays a critical role in change in unbalance of reaction wheels during vibration. The reaction wheels used for spacecraft applications are balanced to very fine levels such that the on-orbit vibrations generated by a running reaction wheel does not affect the performance of payloads onboard a spacecraft (especially optical payloads). Random vibration tests may slightly increase the unbalance of a reaction wheel, but the change in unbalance of a wheel after vibration test must be limited within specified values. As part of qualification of a new generation reaction wheel it was subjected to random vibration test. After the test, it was seen that the dynamic unbalance value after vibration test has gone out of specification. This study indents to find the reason for this deviation and propose design improvements for limiting the change in dynamic unbalance values within the acceptable range.

Keywords Reaction wheel · Nut factor · Fastener joint

D. Syamdas (✉) · K. Pradeep · J. J. Chaman · K. Anilkumar
ISRO Inertial Systems Unit, Thiruvananthapuram 695013, India
e-mail: d_syamdas@vssc.gov.in

© The Author(s), under exclusive license to Springer Nature Singapore Pte Ltd. 2023

R.S. Priyadarshini and T. Sundararajan (eds.), *Advances in Small Satellite Technologies*, Lecture Notes in Mechanical Engineering, https://doi.org/10.1007/978-981-19-7474-8_19

1 Introduction

All components of a launch vehicle or spacecraft are subjected to various kinds of loads during the ascent phase. Low-frequency loads are generated by events like engine shut off/ignition, wind gusts, and quasi-static loads. Other environments are acoustics, random vibration, sine vibration, and shock. Acoustics and random vibrations are generated during aerodynamic events like buffeting and during lift-off phase. Shock loads are generated as a result of separation events like separation of stages, payload fairing, spacecraft, etc. [5]. Sine vibrations are generated mainly due to pogo and slosh in liquid fuel and/or oxidizer tankages. The frequency band of all these events is so wide that it is practically impossible to design a spacecraft component without resonating at any of its natural frequencies.

Reaction wheels for spacecraft are subjected to qualification level random vibration tests. These tests are done to establish the design margins and to bring out workmanship-related issues. During this test, they are subjected to random vibration typically for 120 s in each of the three principal axes. During random vibration environment, all the structural frequencies (fundamental resonant frequencies and higher harmonics) within 20–2000 Hz will be excited simultaneously. The excited range of frequencies is significantly wide, and it is practically impossible design sub-systems without any natural frequencies within the range of excited frequencies. Hence, they must be designed in such that they have to perform its intended function satisfactorily during these tests, even when they are resonated at their natural frequencies. Flywheels of reaction wheels are subjected to very high vibration loads; these loads are transferred to bearing unit through the assembly joints. The integrity of assembly joints plays a critical role in the change in unbalance of a reaction wheel after vibration. The reaction wheels used for spacecraft applications are balanced to very fine levels such that the on-orbit vibrations generated by a running reaction wheel does not affect the performance of payloads, especially optical payloads onboard a spacecraft. Random vibration tests may slightly increase the unbalance of a reaction wheel, but the change in unbalance of a wheel after vibration test must be limited within specified values.

As part of qualification of a new generation reaction wheel it was subjected to random vibration test of 11.77g_RMS along the spin axis. After the test, it was seen that the dynamic unbalance value after vibration test was 18 g-cm2 as against the specified value of 8 g-cm2. This study indents to find the reason for this deviation and propose design improvements for limiting the change in dynamic unbalance values within the specified limits. In order to get better insight into the joint design, it was necessary to estimate the slipping margin, clamping load, and the nut factor.

Nut factor of a preloaded fastener joint is one of the important properties that determines the magnitude of the clamping force for a given tightening torque. Nut factor and diameter of the fastener establishes the torque preload relationship through [2, 4]

$$T = Kfd \tag{1}$$

where T is the applied torque, f is the preload, and d is the nominal diameter of the screw. The value of nut factor depends on various factors like friction and tolerance between the thread, the kind of thread, the surface condition of the object being clamped, etc. While performing design computations, if the assumed nut factor is higher than the actual value, then the estimated clamping load would be lower and hence would give a conservative estimate on the slipping margins of the joint, but it would underestimate the stresses on the fastener. If the assumed nut factor is lower than actual, it would be vice versa. Hence, it is very important to use the right value of nut factor in the design computations for a fastener joint.

Due to the complex nature of the nut factor and the factors influencing the value of it, the ideal method for evaluating it is by experimental means. This study was done as part of preload characterization and nut factor estimation of the bolted joint between bearing unit housing and flywheel of reaction wheel. The flywheel is fastened to the bearing unit using six M4 × 30 mm, standard 12.9 class bolts. The bolts were torqued with 4.5 Nm torque. Exact value of nut factor of the joint was not available for computation of the fastener preload on the joint; therefore, as part of the characterization, a test setup was made for this purpose.

Nut factor estimation studies are generally done by placing a load washer in between the screw joint and monitoring the compressive strains on the load washer for applied torque levels. From the measured compressive strain, the fastener preload force is computed using the load-strain calibration data of the load washer. The bearing unit housing flange has a seating area of 8 mm for the flywheel-bearing unit interface. Standard load washers available commercially for preload characterization studies could not be accommodated in this very limited space. Hence, as part of measuring the preload on the joint, cylindrical-shaped load washers with an outer diameter of 7.9 mm, inner diameter of 4.1 mm, and length of 10 mm was fabricated and used for the test (see Figs. 1 and 2).

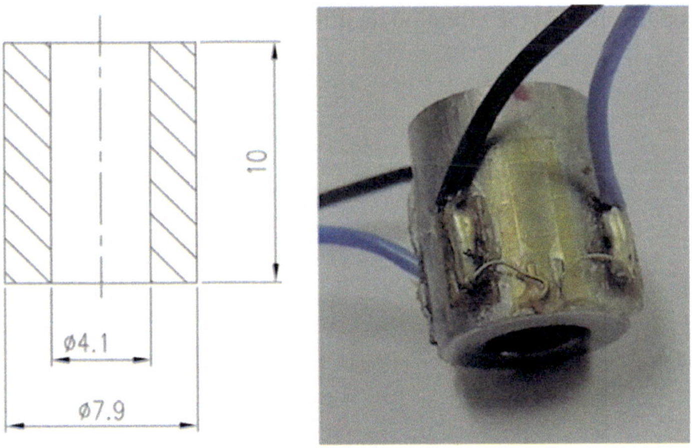

Fig. 1 Dimensions of the load washer and load washer with strain gages on diametrically opposite locations

Fig. 2 Load washer under calibration in UTM

Two strain gages (strain gages 1 and 2) were instrumented on diametrically opposite positions on the load washers (Fig. 1). Compressive load up to 10000 N was applied on the load washers, and the compressive strains were monitored (see Figs. 3 and 4 for Load v/s strain data). The factor of safety the load washer would have at 10 kN force is 1.6. The calibration data of the load washers showed no significant nonlinear behavior. This calibration test was carried out on a universal testing machine. This calibration (load v/s strain) data of the instrumented load washers was used for estimating the joint preload force, preload scattering factor, and nut factor.

As part of the study, fasteners with different materials and/or surface coatings were tested for estimating the nut factor, joint preload, and preload scattering factor. These values were used to compute the slipping margin for the fastener joint. Based on these computations, necessary improvements were implemented, and the improved design has successfully cleared the vibration qualification test.

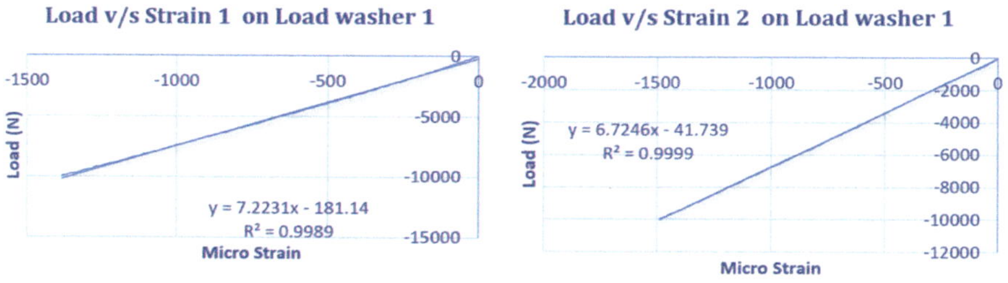

Fig. 3 Load strain calibration data of strain gages 1 and 2 of load washer 1

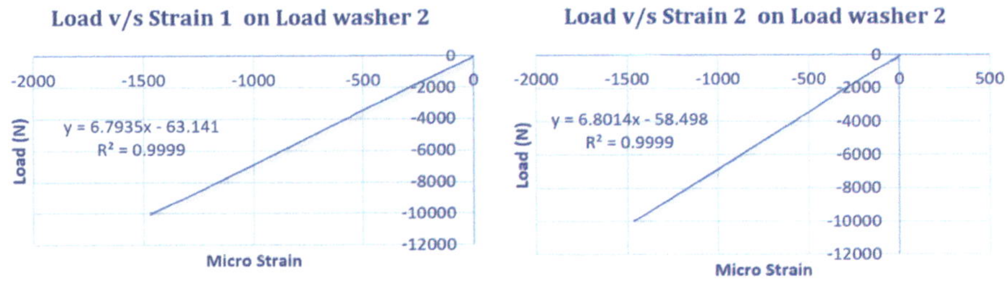

Fig. 4 Load strain calibration data of Strain gages 1 and 2 of Load washer 2

2 Load Washer Calibration Results

The load washers were calibrated multiple times in order to check for the repeatability and linearity of the load washer. It was seen that the response of the load washers was highly linear and repeatable. A straight-line fit was generated for the calibration data. This data was used to measure the clamping load at the fastener joint.

The load-strain data shows very good linearity, and hence this method of testing is expected to give a good estimate of the preload on the fastener. Force computed using this line fit is averaged to minimize the errors and to get a better estimate of the preload force on the joint/fastener.

3 Experimental Procedure

The flywheel-bearing unit interface is joined using six M4 bolts. The six mounting points were used to get six sets of readings. During the test, instrumented load washers were used to measure the clamping load at the joint. The compressive strains (1 and 2) on the load washers were monitored. The value of strain was used to estimate the clamping load on the joint using load washer calibration data (Fig. 5).

The test was carried out with four types of M4 screws, viz., (Table 1)

1. Cadmium coated Ti6Al4V screws
2. Standard 12.9 class black finish screws
3. Al I.V.D. + MoS2 Coated 12.9 class screws
4. Al I.V.D. + MoS2 Coated E40CDV20 screws

4 Design Calculations for Bolted Interface

Tightening a group of fasteners to a predetermined torque value does not ensure the same preload on every fastener location. Geometric, material and surface variation will introduce some amount of scatter. Accounting the value of scatter in the design

Fig. 5 Bearing unit housing used for test

Table 1 Nut factor mean and standard deviation for different types of screws

Type of screw	Nut factor (Mean)	Nut factor (Std. Deviation)
Cd coated Ti6Al4V	0.21	0.02
Std. 12.9 class black finish	0.22	0.02
Al IVD + MoS2 coated 12.9 class	0.13	0.005
Al I.V.D. + MoS2 Coated E40CDV20	0.14	0.01

computations would give a better estimate on slipping margin of the joint and the margin of safety during tightening of the bolt. The scatter in preload δ is also estimated during the test discussed in previous section.

Various bolt parameters used for design calculations are:

d	nominal diameter of screw
p	pitch of screw
μ_v	Thread friction and collar friction of the bolt
μ	Joint friction coefficient
d_s	Diameter at stress cross section
A_s	Stress cross section of bolt thread
λ_a	Ratio of external load carried by the assembly
λ_s	Ratio of external load carried by the screw
P_{min}	Minimum pre-load of the screw
P_{max}	Maximum preload of the screw
P_{nom}	Nominal preload of the screw
C_t	Torsion torque

σ_{yield}	Yield strength of screw material
σ_{ultimate}	Ultimate strength of screw material
δ	Preload scattering factor
F_{axial}	Out-of-plane external force
F_{shear}	In-plane external force
τ	Shear stress
σ	Axial stress
$\sigma_{\text{von mises}}$	*Von mises stress*
T	*Bolt tightening torque*
SM_{slipping}	Safety margin w.r.t slipping of the assembly.

Equations used for design calculation are [1]

$$P_{\text{nom}} = \frac{P_{\max} + P_{\min}}{2} \tag{2}$$

$$\delta = \frac{P_{\max}}{P_{\min}} \tag{3}$$

$$C_t = (0.161 \times p + 0.583 \times \mu_v \times d_2) \times P_{\max} \text{ where } d_2 = d - 0.6495 \times p \tag{4}$$

$$T = \left(0.161 \times p + \mu_v \times \left(0.583 \times d_2 + \frac{d_m}{2}\right)\right) \times P_{\text{nom}} \tag{5}$$

$$\text{where } d_m = \frac{2}{3} \times \frac{d_{\text{ext}}^3 - d_{\text{int}}^3}{d_{\text{ext}}^2 - d_{\text{int}}^2} \tag{6}$$

$$SM_{\text{slipping}} = \frac{\mu \times P_{\min} - \lambda_a \times SF_{\text{slipping}} \times F_{\text{axial}}}{SF_{\text{slipping}} \times F_{\text{shear}}} - 1 \tag{7}$$

Safety margin with respect to slipping of the joint is computed using minimum preload (see Eq. 7); this gives a conservative estimate of the quantity.

Shear stress in screw due to tightening torque

$$\tau = \frac{16 \times C_t}{\pi \times d_s^3} \tag{8}$$

Von Mises stress in the screw [4]

$$\sigma_{\text{von mises}} = \sqrt{\sigma^2 + 3 \times \tau^2} \tag{9}$$

Safety margin w.r.t tightening of screw:

$$SM_{\text{tightening}} = \frac{\sigma_{yield}}{\sigma_{\text{von mises}}} - 1 \tag{10}$$

When a fastener tightened with a tightening device such as a torque wrench is exposed to simultaneous torsion and tension, will yield at a slightly lower level of tensile stress than a fastener subjected to pure tension. This same fastener, however, will support a higher tensile stress in service before yielding further because the torsion stress component will, in general, disappear rather rapidly after initial tightening in most cases. The shear stress due to torsion is created by frictional force developed between the mating threads of the fastener joint; therefore, the magnitude of it is proportional to the thread friction, and hence the torsional stress component for a lubricated screw is relatively low. The torsion stress will consume a part of the total strength of the bolt only while the torsion stress is present. Once the torque wrench is removed, the torsion stress will tend to disappear, and the fastener will recover its full tensile capacity after the torsion stress has disappeared [3]. The tensile stress required to further yield the fastener is higher than that required to yield, while it is being tightened.

5 Results and Discussions

The preload characterization and nut factor estimation test had given a better insight into the behavior of the fastener joint. A better estimate of clamping load, preload scattering factor, and mean and standard deviation of nut factor for various bolt types were obtained. Bolts used in aerospace application which are not reused could be preloaded up to 90% percentage of the allowable stress. However, torsional stress induced during the tightening process of the bolt must be estimated, and positive safety margin during tightening of the bolt should be ensured during tightening (Table 2).

MoS2-coated E40CDV20 screws were used for the reaction wheel assembly with a safe tightening torque of 4.6 Nm, this would induce a nominal clamping load of 8212 N at the fastener joint. This clamping load would provide necessary margin of safety with respect to slipping even under worst-case situations. This assembly was subjected to qualification level random vibration tests and the dynamic unbalance of the reaction wheel was within the required specification of 8 g-cm2.

Table 2 Slipping margins for different types for screws for maximum possible tightening torque

	Std. black finish 12.9 class	MoS_2 coated 12.9	MoS_2 coated E40CDV20	Cd coated Ti6Al4V
Yield strength (MPa)	1100	1100	1300	850
Nominal axial preload (N)	5717	7155	8212	4319
Margin of safety w.r.t slipping	−0.35	−0.03	+0.11	−0.53

6 Conclusions

The increase in unbalance of the reaction wheel was found to be triggered due to slipping occurring at the flywheel-bearing unit joint. This was found to be due to insufficient clamping force at the joint. Higher clamping force was applied at the fastener joint, and positive margin of safety for slipping of the joint and yield of bolt was ensured. The reaction wheel assembly with higher clamping load at the fastener interface was subjected to vibration test, and the dynamic unbalance of the reaction wheel was within the required specification. As an improvement for the future, the joint would be redesigned with larger diameter screws which would increase the clamping load on the joint in order to improve the integrity of the joint further.

Acknowledgements This research was supported by the Scientist/Engineers of CCTD/CMSE and SASD/IISU at their facility. Their sincere support during the work is gratefully acknowledged.

References

1. Astrium (2003) Bolted interface calculation
2. Barrett RT (1990) Fastener design manual. NASA, Cleveland, Ohio
3. Bickford JH (2008) Introduction to the design and behaviour of bolted joints. CRC Press, New York
4. Shigley JE (2006) Mechanical engineering design. McGraw Hill, New York
5. Steinberg DS (1988) Vibration analysis for electronic equipment. Wiley, New York

Design and Development of a 3U CubeSat for In-Situ Radiation Measurements

T. K. Anant Kumar, Ishan Sarvaiya, E. Harshavardhan, Mallikarjun Kompella, Dhruva Anantha Datta, Ch. Sai Abhishek, Vivek Garg, Yugal Joshi, Parthasarthi Samanta, Sai Krishna Prasad, S. Priyadarshini, Priyanshu Jain, S. V. Janakiram, K. Surya Sudhakar, Sandeep Prasad Shaw, Harris V. John, Divyansh Prakash, Devashish Bhalla, Atharva Kulkarni, Sankalp Vishnoi, Sivaranjini, Aman Naveen Murala, Joji John Varghese, P. Suhail, H. Priyadarshan, M. S. Harsha Simha, Sudharshan Kaarthik, V. S. Sooraj, P. Raveendranath, C. R. Bijudas, Umesh R. Kadhane, and P. Chakravarthy

Abstract Understanding the radiation environment of the low earth orbit (LEO) is of considerable scientific interest, due to its unpredictable nature and variability. Temporal and spatial distribution of the ionizing as well as non-ionizing radiation levels in the LEO are especially important considering the negative impact these radiation have on the spacecraft avionics by causing Single Event Upsets (SEUs) and single-event latchups (SELs). With in-situ data about the radiation environment, we can gain an insight on the reliability of various Commercial off-the-shelf (COTS) in the space environment, as ground-based testing with simulated radiation levels for a long duration is often not practical. Further, the data collected about the ionizing radiation levels can also prove useful for the upcoming human space flight mission planning, wherein shielding against ionizing radiation is an important aspect to consider regarding the safety of the astronaut. With this primary objective of in-situ radiation dosimetry, the Ahan 3U CubeSat has been developed with a mass of around 2 kg and a nominal power consumption of less than 3 Watts. The satellite contains two different sensors, a Geiger-Muller Counter and a Radiation Field Effect Transistor (RADFET) to monitor radiation dosage in the LEO. This paper describes the design and development of the above-mentioned CubeSat mission.

Keywords Radiation dosimetry · Geiger muller counter · Radiation Field Effect Transistor (RADFET)

T. K. Anant Kumar (✉) · I. Sarvaiya · E. Harshavardhan · M. Kompella · D. A. Datta · Ch. Sai Abhishek · V. Garg · Y. Joshi · P. Samanta · S. K. Prasad · S. Priyadarshini · P. Jain · S. V. Janakiram · K. Surya Sudhakar · S. P. Shaw · H. V. John · D. Prakash · D. Bhalla · A. Kulkarni · S. Vishnoi · Sivaranjini · A. N. Murala · J. J. Varghese · P. Suhail · H. Priyadarshan · M. S. H. Simha · S. Kaarthik · V. S. Sooraj · P. Raveendranath · C. R. Bijudas · U. R. Kadhane · P. Chakravarthy
Indian Institute of Space Science and Technology, Thiruvananthapuram 695547, India
e-mail: anant.telikicherla@ualberta.ca

H. Priyadarshan
e-mail: priyadarshnam@iist.ac.in

© The Author(s), under exclusive license to Springer Nature Singapore Pte Ltd. 2023
R.S. Priyadarshini and T. Sundararajan (eds.), *Advances in Small Satellite Technologies*, Lecture Notes in Mechanical Engineering, https://doi.org/10.1007/978-981-19-7474-8_20

1 Introduction

The magnetic fields surrounding the earth and the earth's rotation about its axis give rise to highly dense radiation zones called the Van Allen belts. The radiation particles in these belts seem to originate from the solar winds which are eventually captured by the Earth's magnetic fields [1]. Conventionally, the inner Van Allen belt lies in an altitude range of 1000 to 12,000 km. However, in certain geographical areas, the inner boundary may decline down to roughly 200 kms. The South Atlantic Anomaly (SAA) is one such region, located at an altitude of approximately 500 km, spanning from $-50°$ to $0°$ geographic latitude and from $-90°$ to $40°$ longitude [2]. Such regions are characterized by unexpected events which completely change the orientation of the Van Allen belts; one such example being the formation of a narrow third belt of charged particles in September 2012 [3]. The primary motivation of our mission is that the unpredictable nature and continuous transformation of the radiation environment makes real-time monitoring and in-situ measurements a necessity. The current acclaimed standard models, namely, AP-8 and AE-8 (used for simulating trapped proton and trapped electron flux) have inaccuracies when modelling the fluctuations in radiation levels during solar maximum and minimum conditions [4]. The data from the proposed mission can be used to understand the temporal changes in the radiation levels by comparison with previously collected data by missions such as Hiscock Radiation Belt Explorer (HRBE) [5] and Cat-1 [6].

The Ahan mission is IIST's first student satellite mission developed at Small-spacecraft Systems and PAyload CEntre (SSPACE) of IIST, which is highlighted in the name Ahan derived from a Sanskrit word meaning 'the first ray of the sun'. The Ahan satellite is built with the objective of monitoring the space radiation environment using new, compact, and robust disruptive small satellite technologies. Most subsystems of Ahan are indigenously built at IIST, the details of which are summarized in Table 1. A successful demonstration will show the feasibility of radiation environment monitoring using small satellites and can lead to implementation of a future constellation of small satellites that continuously monitor the radiation environment. In addition to the scientific objective of in-situ radiation dosimetry, the Ahan mission is also aimed as a technological demonstration of the various in-house developed small satellite subsystems. Correlating the radiation data with the on-orbit performance of the subsystems will lead to a better understanding of the negative impact these radiations have on the spacecraft avionics by causing single event upsets and single event latch-ups.

2 Science Objectives and Payload

In order to understand the amount of radiation the Ahan satellite will be exposed to in orbit, an analysis was conducted to predict the expected incoming flux and total

Table 1 Ahan specifications

Dimensions and mass	34 × 10 × 10, Standard 3U form factor, 2 kg (approx.)
Power	3 W (Nominal), 8 W (during RF transmission)
Proposed orbit	500 km, 97.8 Inclination
Science objective	In-situ radiation measurement in the LEO
Engineering objective	To flight qualify in-house developed small satellite subsystems
ADCS	Passive Magnetic Attitude Control. Uses a 0.55 Am^2 magnet and hysteresis rods
CDH	Microsemi SmartFusion2 SoC, 128 GB SD card, 64 MB Flash
EPS	2S2P battery pack (6–8.4 V), 4 body mounted panels, Load switches, VC sensors
Comms	UHF transceiver (NanoAvionics), 9600 bps, half-duplex

dose rate of different types of ionising radiations. These include solar energetic particles (SEPs), galactic cosmic rays (GCRs), trapped protons, and trapped electrons. Considering a tentative launch date of September 2021, the cumulative radiation flux over a mission duration of 6 months was analysed using the SPENVIS software [7] for different altitudes ranging from 350 to 900 km. Figure 1a,b show the cumulative proton flux estimated at 500 km and 700 km, respectively, using the standard AP-8 model that covers an energy range of 0.1 to 400 meV for trapped protons. Similarly, Fig. 1d,e show electron flux estimated at 500 km and 700 km, respectively, using the standard AE-8 model which covers an energy range of 0.04 to 7 meV for trapped electrons. Similar plots were also obtained for the range of altitudes, and Fig. 1c,f show the trend of variation of the integral flux in this altitude range. The simulation results show that the energy of protons ranges between 0.1 meV and 300 meV which is about 100 times more energetic than the energy range of electrons (4e-2 to 7 meV). Protons have energy dominance in this region; however, the overall average integral flux of electrons (6.8596e + 4 for 500 km) is higher than that of protons (8.0822e + 2 for 500 km). Further, it is also observed that the integral flux increases exponentially with altitude, which can cause spacecraft avionics failures (due to SEUs and SELs) at higher altitudes. Thus, following the simulation, an altitude of 500 km was considered most suitable as it provides decent radiation exposure without saturating the radiation sensor/ harming the spacecraft avionics. It is also observed that the regions of higher radiation flux are located near the south polar region, in accordance with the South Atlantic Anomaly. Based on the results of the proton and electron flux, a high inclination polar orbit (97.8°) that passes through the South Atlantic Anomaly region was chosen for further analysis. The incoming flux of galactic cosmic rays in this orbit was also analysed using different models which indicate that particle flux having a very wide spectrum of energies will be observed in such an orbit. The details of the analysis are available in the internal technical report.

To characterize the radiation, the science payload of the Ahan mission consists of two sensors: a Geiger-Muller Counter (GMC) and a RADFET-based radiation sensor which take two different measurements. The GMC estimates average ion density by

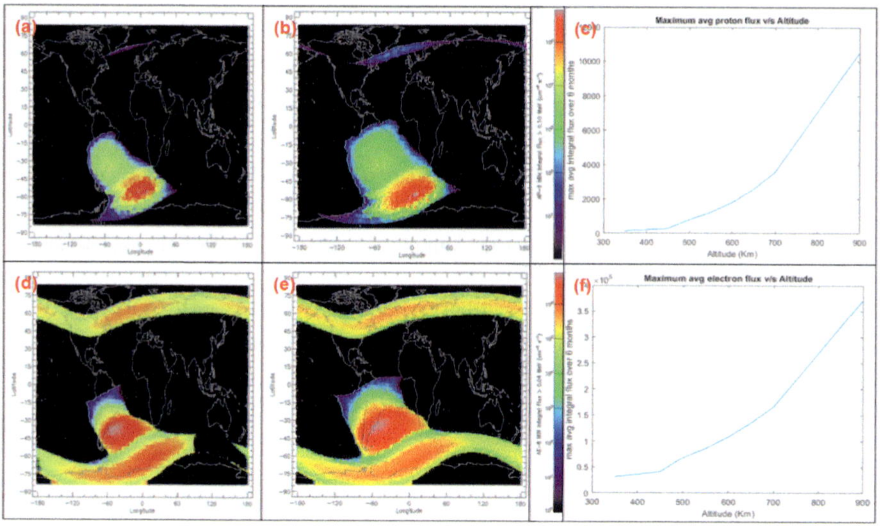

Fig. 1 **a** Proton Flux at 500 km, **b** Proton Flux at 700 km, **c** Variation of Proton Flux with Altitude. **d** Electron Flux at 500 km, **e** Electron Flux at 700 km, **f** Variation of Electron Flux with Altitude

counting the amount of incident radiation received by the satellite. The payload design consists of a commercially available end window GM detector with a gas filling of Ne + HaI, typically used for Beta and Gamma ray counting applications. The detector has an outer diameter of 14.5 mm, a sensitive length of 39 mm, and requires a 500 V supply voltage to operate. An engineering model has been developed that contains the Geiger-Muller tube, as well as the power supply circuit and a readout circuit. For radiation dose measurement, the Ahan mission utilizes novel low-cost technology to conduct the radiation dosage measurement using a Radiation Field Effect Transistor (RADFET). The RADFET can be used to measure a cumulative dose using a nonlinear and a pre-recorded calibration curve. An engineering model has been designed in house in order to implement the read-out circuit for the RADFET using a high-resolution ADC. The science traceability matrix in Table 2 summarizes the science objectives and the instruments used to achieve them.

Table 2 Science traceability matrix

Science objective	Measurement	Instrument
Average density of ionizing radiation within the South Atlantic Anomaly (SAA) region	Ion density	GMC
Radiation exposure rates on spacecraft avionics	Ion density	GMC
Cumulative radiation dose received by spacecraft avionics in the LEO	Radiation Dose	RADFET

3 Mission Design and Analysis

Ahan being a small satellite, has to be launched as a secondary (piggyback) small satellite payload on a PSLV. Thus, the orbital requirement of the primary satellite payload puts a constraint on the orbit selection for the Ahan mission. Typically, most primary satellites launched by the PSLV are delivered to sun-synchronous orbits with altitude ranging from 500 to700 Km. The previous section described the radiation flux observed at different altitudes, which shows that the radiation levels increase significantly as the altitude increases beyond 500 km. Thus, for the Ahan mission, a 500 km altitude was chosen that provides a suitable radiation environment to characterize, which isn't too high to harm the spacecraft avionics. Further, using a polar 500 km orbit enables greater global access as compared to a low inclination orbit, allowing the Ahan mission to create a radiation map across a wider range of latitudes. A high inclination polar orbit also passes through the South Atlantic Anomaly region, which would allow the mission to characterize the unique radiation phenomena in this region. In addition, a sun-synchronous polar orbit gives an added benefit of having lesser eclipse time, which aids the power subsystem of the small satellite. In order to ensure that the satellite remains power positive throughout the mission, a power analysis was performed to simulate the power generation and the state of charge of the battery as the satellite moves in the orbit. Further, an access time analysis was conducted in order to determine the average number of the satellite with the ground station as well as the amount of total data that can be downlinked per orbit. The details of both these analyses are described in the subsequent subsections.

3.1 Power and Detumble Analysis

The Ahan satellite contains a passive magnetic attitude control (PMAC) that aligns the satellite along the earth's magnetic field and can only reduce it's angular rates to 2 degrees/second. Thus, the angle between the sun-vector and the normal vectors of the four solar panels changes with time, which leads to different values of power generated by the satellite. A simulation testbed created in MATLAB that uses an orbit propagator to determine satellite position and also determines the angular rates by solving the dynamics of the satellite. Figure 2a shows the variation of the satellite angular rates after deployment from the launch vehicle considering a worst-case tip-off angular rate of 5°/s in each axis. These are then reduced to less than 2°/s within a time period of 10 days using the PMAC system. The power generation is estimated by using a sun vector model and determining the angle between the sun vector and the solar panels. As shown in Fig. 2b, the satellite remains power positive, the SoC remains between 95 and 100% dropping to the lower boundary in the eclipse.

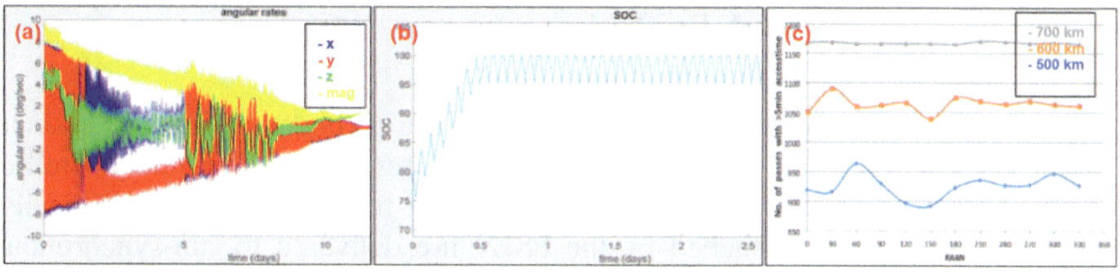

Fig. 2 **a** Satellite angular rates, **b** State of Charge (SoC) of the battery, **c** Access time analysis for different initial RAANs

3.2 Access Time Analysis

An access time analysis was also conducted (using the Systems Tool Kit student version software) that showed that the satellite had an average access time of 10 min per day with the primary groundstation established at IIST. The Ahan satellite contains a UHF transceiver having a data rate of 9600 bps using which an average of 720 KB of data can be downlinked per day. Preliminary estimates on the total data generated by the satellite indicate that less than 10 KB of data will be generated per day, giving a sufficiently large data margin. The graph in Fig. 2c shows access time analysis for different initial RAANs of the satellite. For this analysis, links with access time greater than 5 min are considered, as in a worst-case scenario it will take 1 min for connection establishment, 3 min for downlinking data, and 1 min for uplinking any commands required.

4 Satellite Subsystems Overview

The Ahan satellite is designed with a 3U form factor. A 3D render of the satellite (Fig. 3a) shows how the different components are placed in the satellite. The satellite has dimensions of approximately 34 × 10 × 10 cm and a mass of around 2 kg. Figure 3b shows a system-level block diagram of the entire satellite highlighting the interconnections between the different subsystems. The various subsystems PCB which includes the payload, the on-board computer (OBC), the electrical power system (EPS), and the UHF transceiver are stacked in the middle part of the satellite and connected using the PC104 interface.

The on-board computer is built in-house using the SmartFusion2 M2S090 SoC (System on Chip) FPGA which incorporates ARM Cortex M3 microprocessor, FPGA fabric, and other features on a single chip. The OBC designed for the Ahan mission is an improved version of the previously designed OBCs at IIST (developed for INSPIRESat-1 [8] mission). The OBC is also responsible for running the flight software for the mission which is developed in C language. The communication system of Ahan uses a COTS UHF half-duplex transceiver developed by NanoAvionics with a proposed operating frequency of 437 MHz. A tape-measure monopole antenna that

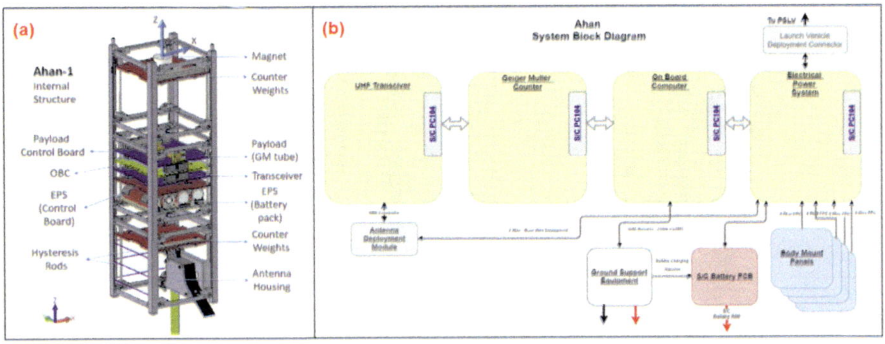

Fig. 3 **a** Ahan 3D CAD model, **b** Ahan system block diagram

is housed in an antenna deployments module (placed near the bottom of the satellite ($-z$ face)) is used for transmission and reception. The antenna is deployed using a burn-wire deployment mechanism that derives heritage from the INSPIRESat-1 mission. The electrical power system of the Ahan satellite consists of a charger board, a battery board, and four solar panels (with 7 solar cells per panel). The battery board consists of lithium ion cells in a 2S-2P configuration, generating a nominal battery voltage of 7.2 V and a nominal battery capacity of 5000 mAh. Since the radiation dosimetry payload doesn't have any stringent pointing requirements, there is no necessity to use an active attitude determination and control system for the Ahan mission. Thus, a passive magnetic attitude control (PMAC) system was used, which contains a permanent magnet with a magnetic moment of 0.55 Am2 (derived mathematical relations developed in previous small satellite missions [9]) is fixed on the z axis of the satellite. Hysteresis rods are added to along the X and Y axes of the satellite to dampen the oscillatory motion of the permanent magnet about the local magnetic field vector. Satellite attitude and angular rates are determined to verify the proper functioning of the PMAC system. This is done using an inertial measurement unit (IMU) sensor present on the OBC that consists of an accelerometer, gyroscope, and a magnetometer. Further, coarse sun sensors are also added to the six faces of the satellite for sun vector determination that can be used to determine the attitude of the satellite when combined with other measurements from the IMU.

The structural design of Ahan is based on the ISIS standard 3U design was developed with minor modifications, so that it can interface with the INLS deployer (developed by VSSC-MVIT). A finite element method (FEM) analysis was done on the structure to analyse the effect of the launch loads as prescribed by the PSLV team. The structure was analysed considering it to be made of aluminium-6061. Static, modal, sine vibration, and random vibration analysis were completed on the satellite structure to make sure that the satellite is able to survive the requirements of the launch vehicle (PSLV). The modal analysis shows a minimum modal resonant frequency of 518.17 Hz, the minimum natural frequency prescribed by ISRO being 90 Hz. A representative displacement contour for the first mode is shown in Fig. 4a. Results from the static load analysis indicate a minimum factor of safety of 6.8 (Fig. 4a), whereas

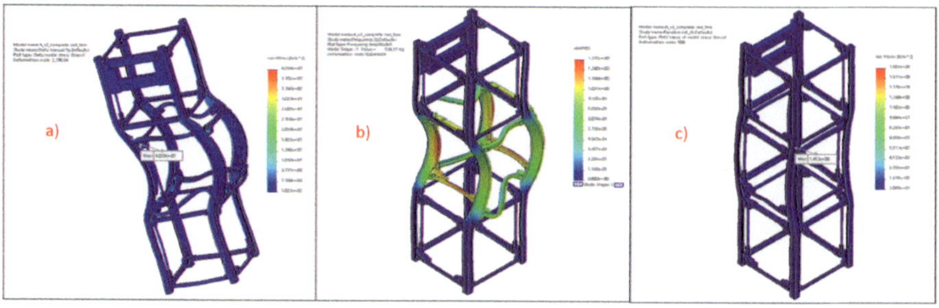

Fig. 4 Ahan structural simulation: **a** Von Mises stress plot—static load analysis in simultaneous application of lateral and longitudinal loads, **b** First mode—representative contour, **c** Random vibration von mises stress plot in X-direction

the FoS from the random vibration analysis is around 1.75 (Fig. 4c). The avionics subsystems as well as payloads of the Ahan mission do not contain any components (such as CCD detectors) having stringent temperature requirements. Thus, the only major requirement of the thermal subsystem is to ensure that the temperatures of different subsystems do not cross beyond the absolute maximum ratings. Moreover, the total power consumption of Ahan is <10 W. It has been observed in small satellites with such low power consumption values that they do not need active thermal control in orbit. Hence, a passive thermal control strategy was adopted. This strategy involves locating effective heat sinks for the satellite and ensuring efficient heat transfer through passive conduction and radiation. This transfer of heat from the PCBs to the skeleton can be achieved using spacers, provided they have good contact with the PCBs. Copper strips can also be placed to enhance the heat removal rate from the PCBs.

5 Integration and Testing

In order to ensure the robust working of the various in-house developed subsystems, engineering models of the subsystems are created and tested, referred to as subsystem wise unit-testing. Once the unit tests are completed, interfacing tests are performed between the subsystems. The different subsystems are then integrated together in the form of a Flat-Sat (which consists of the avionics boards laid out on a bench without the structure) in order to test the functionality of the satellite avionics system. This is followed by the assembly of the engineering model (EM) of the satellite with the structure. After the EM of the satellite is tested, the flight model build can be started.

Currently, the engineering models of the OBC, the communications board as well as the RADFET payload board of the satellite have been developed. Shown in Fig. 5a is the test setup for the UHF transceiver during an end-to-end test with the groundstation at IIST. The transceiver was connected to an antenna and the commands were sent to it using a serial utility on a computer. Test packets were also transmitted

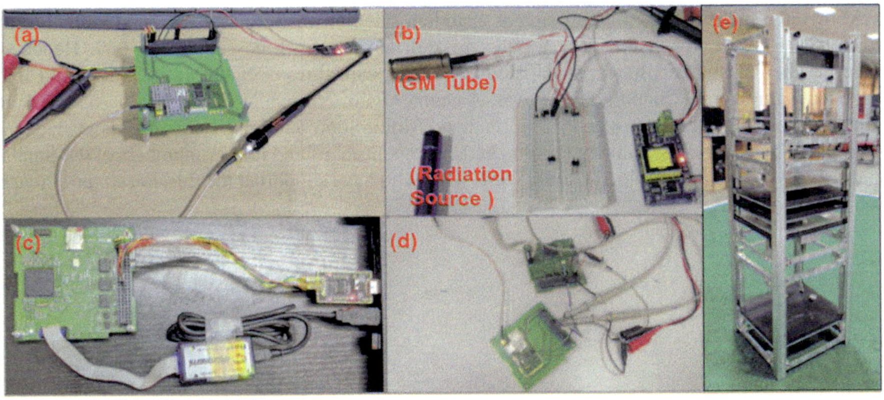

Fig. 5 **a** Transceiver testing, **b** GMC testing, **c** OBC EM testing, **d** UHF-OBC interfacing, **e** Ahan mass dummy

and received in the ground station. Figure 5b shows the breadboard testing of the GMC payload with an external radiation source. Figure 5c shows the unit testing of the OBC board and flight software. During this test, the flight software was run on the OBC, and the data was constantly logged to a computer to ascertain robust functioning of the software scheduler. Once the boards were tested individually, the two subsystems were interfaced together as shown in Fig. 5d. Figure 5e shows the assembled mass dummy model of the Ahan satellite was manufactured in-house at IIST. For the mass dummy model, all the electronic subsystems including the solar panels were replaced with mass dummies made of acrylic, steel, aluminium, etc. The structure is planned to undergo the PSLV-Launch specific qualification tests for different vibration loads.

6 Conclusions

The Ahan small satellite is being developed by the Small-spacecraft Systems and PAyload CEntre (SSPACE) at IIST. The mission will help to demonstrate that in-situ radiation dosimetry can be conducted through low-cost novel disruptive small satellite technologies to enhance the understanding of the radiation environment. For the development of the mission, an analysis has been done to simulate the radiation environment of the LEO region, and also develop various low-cost sensors that can be used to measure and characterize the radiation. Currently, the engineering models of the different subsystems have been developed after multiple iterations of design and testing. The next key steps include flight model fabrication, completing the flat-sat tests, the deployment tests, and environmental tests including thermal vacuum and vibration. After completion of the tests, the Ahan satellite will be delivered for launch to the Indian Space Research Organization in 2021.

Acknowledgements The Ahan mission as a student satellite was formally approved in final form by 2016. However, the main ideas for the mission have been discussed since 2009. Hence, a large number of students have participated in various capacities over the years. We would like to thank Ankit Verma, Sanidhya Vijaywat, Ansuman Palo, Niwashini O, Kaustubh Anand, N R Shankar Ram, Nidish Narayanaa, Prajwal Patnaik, Mohit Soni, Samvram Sahu, Monika Chauhan, Aditya Krishna, Harish S, Pamraat Parmar, and all students/staff and faculty members of the Small-spacecraft Systems and PAyload CEntre (SSPACE) who have supported the mission so far to reach this final form.

References

1. Pierrard V, Rosson GL (2016) The effects of the big storm events in the first half of 2015 on the radiation belts observed by ept/proba-v. Annales Geophysicae, pp 75–84
2. Baker DN et al (2013) A long-lived relativistic electron storage ring embedded in earth's outer van allen belt. Science
3. Cowen R (2013) Ephemeral third ring of radiation makes appearance around earth. Nature
4. Heynderickx D, Kruglanski M, Pierrard V, Lemaire J, Looper MD, Blake JB (1999) A low altitude trapped proton model for solar minimum conditions based on sampex/pet data. IEEE Trans Nucl Sci 46
5. Klimpar DM (2012) Cubesat lessons learned: two launch failures followed by one mission success. In: Proceedings of the 26th annual AIAA/USU small satellite conference, session II: lessons learned, SSC12-II-4, Utah, USA
6. Jove-Casurellas et al (2017) Cat-1 project: a multi-payload cubesat for scientific experiments and technology demonstrators. Eur J Remote Sens
7. Heynderickx D, Quaghebeur B, Speelman E, Daly E (2000) ESA's space environment information system (spenvis)—a www interface to models of the space environment and its effects. In: Proceedings of the 38th aerospace sciences meeting and exhibit, American institute of aeronautics and astronautics, Reston
8. Boyajian S et al (2019) INSPIRESat-1: an ionosphere and solar x-ray observing microsat. In: Proceedings of the 33rd annual AIAA/USU small satellite conference, session V: next on the pad, SSC19-V-06, Utah, USA
9. Gerhardt DT, Palo SE, Passive magnetic attitude control for cubesat spacecrafts. In: Proceedings of the 24th annual AIAA/USU small satellite conference, session VIII: Frank J. Reed student competition, SSC19-V-06, Utah, USA

Performance Evaluation of Aluminum Heatsink for High Reliable CCGA Packages

K. R. Suresha, Santosh Joteppa, and Vinod Chippalkatti

Abstract Digital technology is trending in modern electronic applications. Due to its high reliability, space technology uses electronic packages like CCGA which are molded inside ceramic case hence named ceramic packages. Some of these packages have thermal path from junction (die) to top surface of the package. Thus, external thermal coupling is required to drain out excess of heat to maintain the temperature within allowable limits for safe operation of satellite. Since, these packages are high in cost and one-time programmable devices, cannot be used for R&D purpose; hence, equivalent thermal component which dissipates the way CCGA dissipates is employed to duplicate the model. Aluminum heat sink is used as a thermal coupler to drain out excess of heat. In this paper, effectiveness of heat sink is studied. Thermal analysis has been performed using CFD software by considering different heatsink configurations. Based on analysis results, the best heatsink design is selected and fabricated. Experiments are performed under different boundary conditions as per the analysis. Assembled unit is subjected to thermal testing, and the results are compared with that of thermal simulation, and they are well within the acceptable limit. Using these results, the same heat sink is implemented with actual CCGA to maintain junction temperature within allowable limit specified by the manufacturer.

Keywords CCGA · CFDs · Junction temperature · Heat sink · Thermal coupling

1 Introduction

As technology grows, industries seek advancement in day to day. Electronic industries is one such field spreading its foot print in telecommunication, defense, space,

K. R. Suresha (✉) · S. Joteppa · V. Chippalkatti
Centum Electronic Limited, Bangalore, Karnataka 560106, India
e-mail: sureshak@centumelectronics.com

S. Joteppa
e-mail: santoshj@centumelectronics.com

V. Chippalkatti
e-mail: vinod@centumelectronics.com

© The Author(s), under exclusive license to Springer Nature Singapore Pte Ltd. 2023

R.S. Priyadarshini and T. Sundararajan (eds.), *Advances in Small Satellite Technologies*, Lecture Notes in Mechanical Engineering, https://doi.org/10.1007/978-981-19-7474-8_21

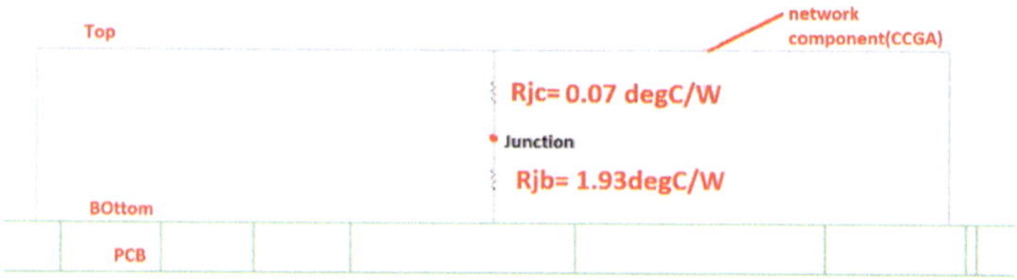

Fig. 1 Network block defines CCGA (thermal resistance from junction to top is less hence heat flow to the top)

medical, etc. As we aware, in space applications, weight is an important constraint which determines the cost of the satellite launch. Advancement in technology replaces bigger modules with small packages with high dissipation and prone to use complex thermal management methods. These small packages construction are such that heat has to drain out through columns or from the top side surface of the component; hence, thermal management plays a crucial role in space electronics for failure-free products.

Digital applications use ceramic packages like CCGA; based on application and load, these packages dissipates more and unable to transfer the heat to board because of high thermal resistance from junction to board. These packages may lead to complete system failure if there is no proper thermal management. While adopting thermal solution, many factors need to be considered such as weight, cost, application or attachment area, availability, acceptability, mode of heat transport, distance from source to sink, power dissipation, etc. Customized aluminum heat sink is designed and used to handle the dissipation of CCGA up to 10 W with 55 °C ambient. Though there are high conductance thermal solutions like heat pipes and cold plates, aluminium heat sink (purely conductive heat transfer) is selected because of its low cost with minimum number of mechanical fixing elements. These heat sink performances are evaluated with analysis using CFD software and experimentation using test setup, which is discussed in Sect. 2.

Figure 1 shows the network block having less thermal resistance from junction to top of the component compared to thermal resistance from junction to bottom of the package. Component junction temperature [1] is controlled by controlling surface temperature of the component when the thermal path is from junction to top surface. Temperature rise from junction to case is depends on the thermal resistance of the component from junction to case. Temperature raise from case to junction is calculated with the formulae

$$Tj - Tc = Rjc * Q$$

where *Tj* is the junction temperature in °C, *Tc* is the case temperature in °C, *Rjc* is the thermal resistance from junction to case in °C/W, *Q* is the heat dissipation in *W*, *Rjb*—thermal resistance from junction to board in °C/W.

2 Thermal Design and Analysis

The practice of usage of simulation software plays a predominant role in reducing time and cost before realizing the product. By creating virtual model and simulating the model gives the areas need to improvise. Heat sinks are simulated with proper boundary conditions to check heat sink adequacy to meet intended specification. The steady state thermal analysis is performed with commercial CFD software. Two concepts have been studied in this work to attain best configuration heat sink [2, 3].

2.1 *Concept 1*

In space application, component temperature is reduced by transferring heat from component to housing through PCB or by introducing external element. Based on this, heat sink is designed and steady-state thermal analysis is carried out [4]. Assembly consists of housing, PCB, HMC (equivalent thermal component), and customized heat sink. Housing made up of Aluminium and HMC is modelled [5] such a way that more heat should flow toward heat sink (Fig. 2).

Heat sink dimensions: Lug diameter—6 mm; Thickness of flat portion—4 mm.

PCB consists of six layers and equivalent thermal conductivity [2] is calculated using copper percentage and layer thickness with help of below formulae.

$$K_{\text{inplane}} = \frac{\sum_{i=1}^{i=n} k_i t_i}{\sum_{i=1}^{i=n} t_i} \ \& \ K_{\text{Outplane}} = \frac{\sum_{i=1}^{i=n} t_i}{\sum_{i=1}^{i=n} \frac{t_i}{k_i}}$$

where K is the thermal conductivity in W/mk; "i" is the Layer number; and t is the thickness of the layer.

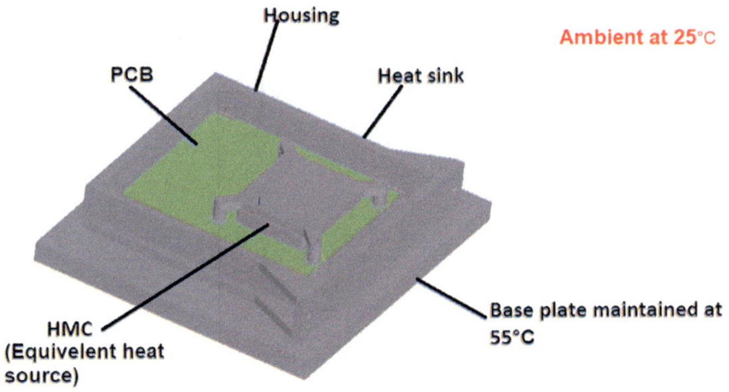

Fig. 2 Concept 1 design with boundary conditions

Table 1 Mesh details from software

Details of mesh			Mesh quality		
Sl. no	Description	Number	Sl. no	Description	Range
1	No. of nodes	1,987,981	1	Face alignment	0.366 to 1
2	No. of elements	1,868,825	2	Volume	9.35e-13 m^3 to 6.11e-5 m^3
			3	Skewness	0.02 to 1

Calculated value of PCB thermal conductivity in in-plane is 39.6 W/mk and out plane is 0.47 W/mk.

Material properties are assigned [1] to each of the components. 8 W heat dissipation is added to the junction of equivalent heat source (HMC) of which almost 7.5 W is drain out from top side of the component. By considering test condition, convection is consider for analysis. Base of the unit is maintained at 55 °C with an ambient of 25 °C. The grid generation plays an important role in the CFD results. Improper grid size has major impact on result. Selection of fine mesh leads to increase in the computational cost and time. Usage of course mesh leads to wrong solution; hence, it is important to take care grid size during mesh generation (Table 1).

The meshed model is simulated using the following set of equations [2].

Continuity Equation

$$\frac{\partial u}{\partial x} + \frac{\partial y}{\partial y} + \frac{\partial w}{\partial x} = 0 \tag{1}$$

Momentum Equation in X direction

$$u\frac{\partial u}{\partial x} + v\frac{\partial u}{\partial y} + w\frac{\partial u}{\partial z} = f_x - \frac{1}{\rho}\frac{\partial p}{\partial x} + \nu\left(\frac{\partial^2 u}{\partial x^2} + \frac{\partial^2 u}{\partial y^2} + \frac{\partial^2 u}{\partial z^2}\right) \tag{2}$$

Momentum Equation in Y direction

$$u\frac{\partial v}{\partial x} + v\frac{\partial v}{\partial y} + w\frac{\partial v}{\partial z} = f_y - \frac{1}{\rho}\frac{\partial p}{\partial y} + \nu\left(\frac{\partial^2 v}{\partial x^2} + \frac{\partial^2 v}{\partial y^2} + \frac{\partial^2 v}{\partial z^2}\right) \tag{3}$$

Momentum Equation in Z direction

$$u\frac{\partial w}{\partial x} + v\frac{\partial w}{\partial y} + w\frac{\partial w}{\partial z} = f_z - \frac{1}{\rho}\frac{\partial p}{\partial z} + \nu\left(\frac{\partial^2 w}{\partial x^2} + \frac{\partial^2 w}{\partial y^2} + \frac{\partial^2 w}{\partial z^2}\right) \tag{4}$$

Energy Equation

$$u\frac{\partial T}{\partial x} + v\frac{\partial T}{\partial y} + w\frac{\partial T}{\partial z} = \alpha\left(\frac{\partial^2 T}{\partial x^2} + \frac{\partial^2 T}{\partial y^2} + \frac{\partial^2 T}{\partial z^2}\right) + \phi \tag{5}$$

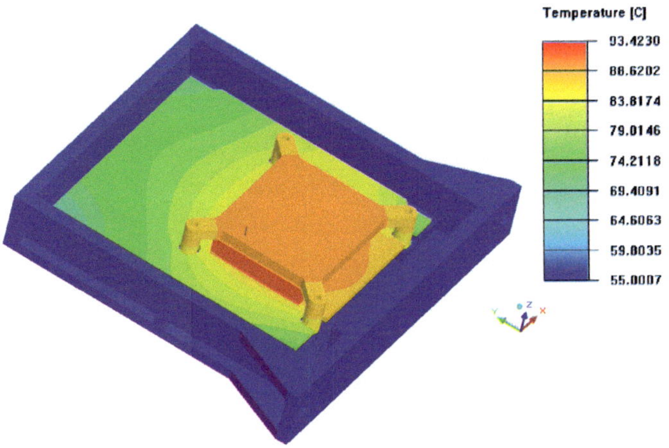

Fig. 3 Temperature profile of assembly (Concept 1)

where (u, v, w) are the component of the fluid velocity and (f_x, f_y, f_z) are body force in direction of (x, y, z) direction, p is pressure, T is temperature, and ϕ is viscous dissipation.

Flow simulation is coupled with thermal analysis, software will calculate heat transfer coefficient h itself and gives the result. The temperature plot is shown below (base plate is not shown) (Fig. 3).

The temperature on equivalent component (HMC) from analysis is found to be 93.4 °C, which is quite high. For further reduction of temperature of the component, heat sink contact area with the housing needs to be increased. Increasing lug size of the heat sink demands more PCB area; hence, the heat sink is projected to housing as discussed in concept 2.

2.2 *Concept 2*

To give more contact area flat portion of the heat sink extended to the housing by proving projection in housing. Contact area got increased and simulation done for the updated thermal model with same configuration as there in concept 1. And mesh parameters are well within the acceptable limit. Thermal model and temperature plots are shown below (Figs. 4 and 5).

The temperature on equivalent component (HMC) from analysis is found to be 83.5 °C. For 55 °C base plate. Concept 2 gives the 10 degree advantage over concept 1 and same heat sink is fabricated for the experimentation.

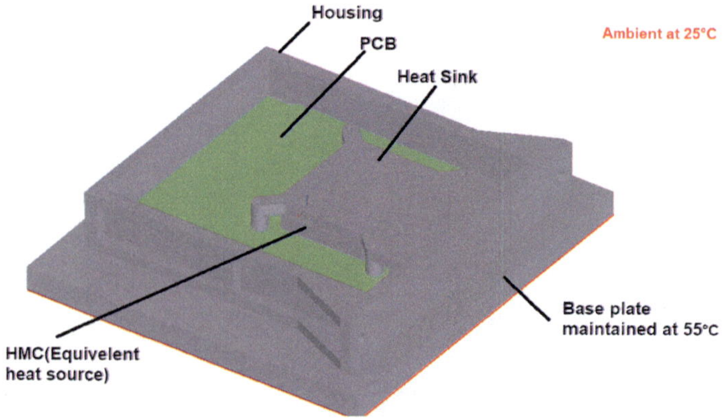

Fig. 4 CAD model (concept2)

Fig. 5 Thermal profile concept 2

3 Experimentation

The experimental set up consisting of PCB with equivalent thermal component (HMC), which is kept inside the aluminum mechanical housing. A 10 Ω thick film resistor (Part no. MP930) is used as a power source, which is attached to the lid of the dummy HMC (equivalent thermal component), and the ends of the resistor are soldered to the HMC pins (For power supply). The package is hermetically sealed. The HMC is raise mounted to minimize the heat flow towards PCB (similar to CCGA). Chotherm of 0.4 mm thickness with thermal conductivity 2.6 W/mk is used as thermal interface material between HMC top and Heatsink. Lambda tdk power supply is used for power up the unit, and K-type thermocouples are used to monitor

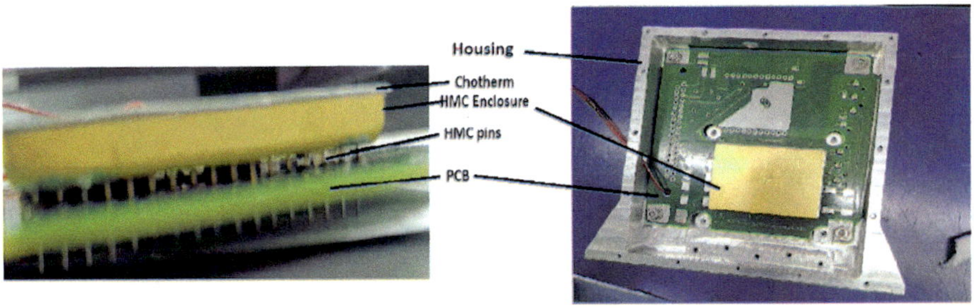

Fig. 6 Experiment test unit

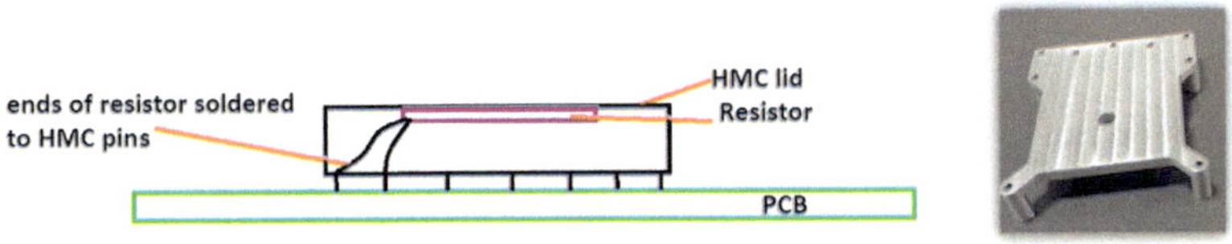

Fig. 7 Equivalent heat source and final heat sink configuration (hole in middle of heat sink drilled for temperature measurement on HMC)

the temperature. All calibrated devices are employed to perform this experimentation [6] (Figs. 6 and 7).

The fabricated heat sink is attached to HMC with thermal interface material chotherm. Hot plate is employed to maintain base temperature at 55 °C. The complete experimental set up is shown in Fig. 8. The resistor is powered by power supply, and the current voltage values are set to dissipate 8 W. Selected locations are monitor for the temperature. Before switching ON the unit, hot plate is allowed to stabilize for 55 °C. After this unit is switched ON, the unit is allowed to stabilize. The temperature values are taken once the unit is stable, and results are recorded. Results are discussed in below section [7].

4 Experimental Results

The unit is tested before fixing the heat sink and the temperature is monitored on the HMC. In experiment, HMC without heat sink reaching 110 °C within 3 min of duration, hence the unit is switched OFF to avoid the damage of resistor. With heatsink integration the temperature on HMC was 83 °C. Which is close to the temperature obtained in analysis (83.5 °C). The temperatures on the unit at selected location are tabulated after stabilization with base at 55 °C (Fig. 9 and Tables 2 and 3).

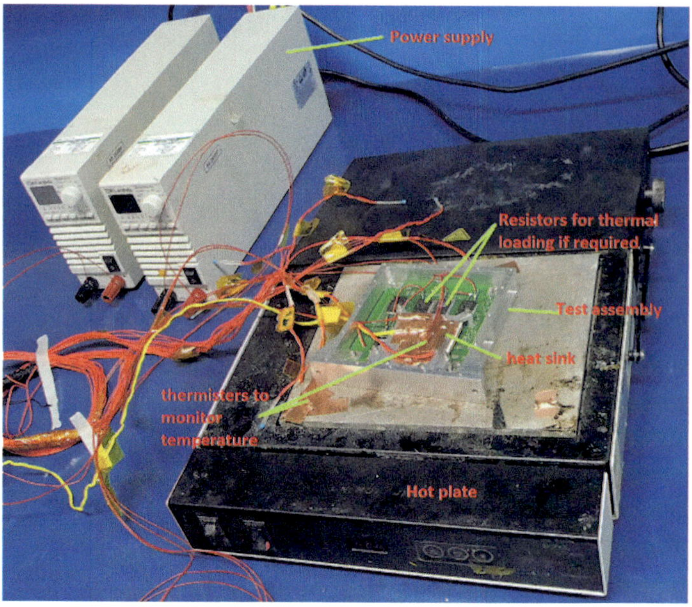

Fig. 8 Experiment test set up

Fig. 9 Temperature monitored locations (resistors on PCB are not used, mounted for additional dissipation to PCB if required)

Table 2 Experimental results

Location	Component	Temperature (°C)
1	HMC top	83
2	Heatsink	68
3	Heatsink	67
4	Heat sink	66
5	PCB	60
6	PCB	53
7	Housing	64

Table 3 Experimental and analysis results comparison

Location	Component	(Experiment) Temperature (°C)	(Analysis) Temperature (°C)	% of error
1	HMC top	83	83.5	0.6
2	Heatsink	68	70	3
3	Heatsink	67	70	4.2
4	Heat sink	66	66	0
5	PCB	60	58	3.3
6	PCB	53	56	5.3
7	Housing	64	62	−3.2

The temperature of the components was crossing 110 °C without heat sink, with the concept 1 the temperature reduced to 93 °C. By projecting the heat sink (concept 2), contact area of the heat sink with housing is increased and it tends to decrease in temperature of the component i.e. temperature got reduce to 83 °C from 93 °C. The results of experimental are compared with analysis results and the maximum deviation is found to be 5.3%, which is well within the acceptable limit.

5 Conclusion

From this work, it is observed that equivalent thermal component (HMC) temperature without any thermal coupling is reaching 110 °C within 3 min, and in order to bring down this high temperature within allowable range heat sinks are employed. Two heatsink concepts have been studied among which concept 2 found to more effective because of more contact area. The temperature values obtained from the experiment are well within the allowable range (max. 5.3% deviation) hence heat sink concept 2 is safe to implement in the real-time application for high-dissipating electronic packages like CCGA and same can be implemented for similar electronic packages whose thermal path is from junction to top with same dissipating ratio.

References

1. Joteppa S, Sharath BK, Chippalkatti V, Analytical and experimental thermal behaviour of an AC–DC convertor with natural convection. HMTC 2019-07-0173
2. Sharath BK, Joteppa S, Dibbi S, Chippalkatti V, Thermal investigation of power supply module(QDR-PSM) for space application using numerical and experimental approach
3. Joteppa S, Singh B, Prasad S, Chippalkatti VS (2012) Analysis and experimental verification of space grade power supply performance under thermo-vacuum environment. Sci Acad Publ Int J Aerospace Sci 1(3):43–48
4. ANSYS Icepak 12.1, user's guide November, 2009
5. Maleki H, Shamsuri AK (2003) Thermal analysis and modeling of a notebook computer battery. Elsevier Sci J Power Sour 115:131–136
6. Lau JH, Yue TG (2009) Thermal management of 3D IC integration with TSV. In: IEEE electronic components and technology conference, pp 635–640
7. Lee RKW, Montero MG, Wright PK (2003) Design methodology for the thermal packaging of hybrid electronic mechanical products. In: ASME, Design Engineering Technical Conferences, DAC-48790, p 110

Navigation and Small Satellites

Big Paradigm Shift in Small Satellite Technology and Applications

Vinod S. Chippalkatti, S. S. Rana, and Rajashekhar C. Biradar

Abstract In recent years, the space sector is going through an incredible transformation. The services from large satellites are actively considered for replacement with smaller satellites which have several advantages compared to the traditional satellites. These are designed and manufactured at a lower cost with flexibility in reducing the lead times. In addition, several low earth orbit satellite Iinternet constellations using small satellites are already being developed for low-cost and low-latency Internet access from space. The services are comparable with addition of the newer functions. The experimentation possibilities using new technologies are more feasible resulting in an accelerated pace of innovation. Starting from the selection of components including those which are rated non-space Commercial Off-The-Shelf (COTS), following the revised derating guidelines, up screening the modules and subsystems to the required environmental stress levels, and complying with the modified product realization and packaging guidelines are some of the path-breaking practices that are becoming common in faster realization of small satellite. Modular designs and application of best practices of a manufacturing industry set up dominate the cost expectations from the small satellite programs. The miniaturization trends in multiple technologies such as in sensors, computing and electronic subsystems have benefited different types of small satellites. Communication, data, earth observation and navigation areas are benefitted in small satellites with the technological advancements in the commercial and industrial domains and their chipsets. While the small satellites are generally in low earth orbits (LEOs), they can offer services comparable to larger satellites in medium earth orbits (MEOs) or geosationary earth orbit (GEOs). This paper describes the small satellite classification, drivers for their growth, technology

V. S. Chippalkatti (✉) · S. S. Rana
Centum Electronics Limited, Bangalore 560106, India
e-mail: vinod@centumelectronics.com

S. S. Rana
e-mail: ssrana@centumelectronics.com

V. S. Chippalkatti · R. C. Biradar
School of ECE, REVA University, Bangalore 560064, India
e-mail: dir.ece@reva.edu.in

© The Author(s), under exclusive license to Springer Nature Singapore Pte Ltd. 2023

R.S. Priyadarshini and T. Sundararajan (eds.), *Advances in Small Satellite Technologies*,
Lecture Notes in Mechanical Engineering, https://doi.org/10.1007/978-981-19-7474-8_22

trends, and applications. It also explains the eco-system available and the players involved in the growth of the small satellite markets.

Keywords Small satellites · Space missions · COTS · Miniaturization · Space economy

1 Introduction

The development and popularization of small satellites represent a shift in the traditional space sector. Small satellite industry is experiencing a revolution in the last few years across all aspects of the space economy. They incorporate the best of technologies from the industrial, aerospace, and to some extent from the commercial world.

There is no standard definition of a small satellite and multiple sources [1] have classified them based on mass, volume, capabilities, and cost. One set of definition based on wet mass can be: Femto satellites (10–100gms), Pico satellites (0.1–1 kg), nanosatellites (1–10 kg), micro-satellites (10–100 kg), and Mini-satellites (100–500 kg). Satellites whose mass is less than 200Kgs are generally discussed under small satellite category.

Small satellites also include CubeSats, a special class whose volume is a cuboid of $10 \times 10 \times 10$ cm and wet mass of <1.33 Kgs. These have the form factors of 1-3U and 6U. A smaller version called "Pocket Cube Standard" defines a satellite with a wet mass of 125 g having a dimension of $5 \times 5 \times 5$ cm. These CubeSats have an additional advantage of being standardized for their containerization, mass production and launcher intefeaces. This ultimately brings increased manufacturing efficiency and cost reduction.

The number of applications and users for such small satellites are increasing rapidly and covers remote sensing (earth observation and imagery-based intelligence), space-based communication (broadband and low latency internet), scientific experimentations, and technology demonstrators. The technological adavcenement are in the areas of synthetic aperture radar, high-resolution optical imaging, onboard processing, hyperspectral imaging, advancements in miniaturization, propulsion systems, and debris management.

2 Drivers for Small Satellites Growth

Flexible and agile design, commercial-off-the-shelf components, digitalization and miniaturisation, dual-use, and spin-in have become synonymous with the ongoing change within small satellite systems. Today's business models [2] thrive not only on

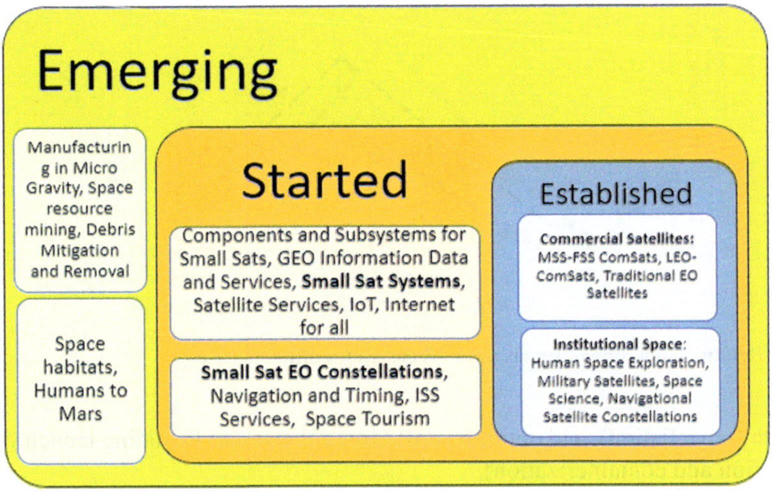

Fig. 1 "Established → Emerging" areas of space economy

technological improvements, but also on shorter generation cycles, aggressive spin-in approaches, and a consequent trade-off between risk, cost, and time to market (see Fig. 1).

The following factors drive the small satellite revolution resulting in disproportionate growth of their technology and applications [3].

- *Latest Technology*: Unlike traditional satellites that use space-qualified materials and components, which have a gap of 10–15 years compared to the latest technologies. The miniaturization of electronics and increase in reliability and performance have enabled small satellites to generally attempt to use COTS components. Use of Micro Electromechanical Systems (MEMS), plug and play systems, Rapid prototyping techniques are increasing the overall speed of hardware realization.
- *Standardization*: The standardization of satellite bus and associated modules makes small satellite technology accessible.
- *Platform for New Technologies Testing*: Small satellites enable cheaper and faster qualification of new subsystems and systems in space environment.
- *Life Cycle Reduction*: Since small satellites have simpler architecture and make use of easily available COTS components, the development phase is shorter.
- *Lower Cost*: By introducing single purpose mission objectives unlike multi-payload missions, the complexity is reduced resulting in faster development and demonstration and cost reduction.

Answers to questions like "What is the right size and cost for small satellites" are attempted by many researchers and reviewers [4]. Figures of Merit to assess the right sizing of the small satellites are evolving. Spacecraft Utility (factor of satellite volume, payload volume, and power), mission utility (aggregate value of a small

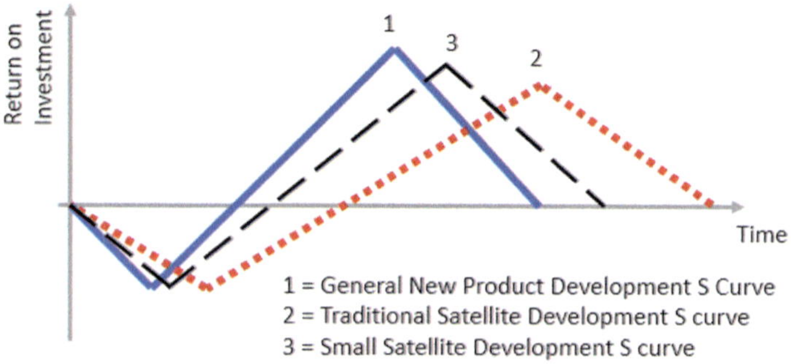

Fig. 2 Return on investment versus time—S curves for different cases

satellite constellation), and optimum cost (break even point including launch vehicle integration and containerization).

The return on investment versus time frame S curves for general product development (see Fig. 2), traditional satellite subsystem, system development, and for the emerging small satellite systems. For traditional space technology, especially on the hardware side, the product development phase is at least 5 years but may take up to 15 years or longer. These extended development phases create a significantly longer "valley of death" that needs to be bridged financially, represented by a longer S-curve (curve-2). Nevertheless, the phase of productivity and harvesting is considerably longer, typically reaching 10–15 years, resulting in a minimum of 30 years between product development and product obsolescence. NewSpace and small satellite system approaches, however, reduce the initial very long research and innovation phase, pushing the space tech S-curve (curve-3) to the left, nearer to the general tech S-curve (curve-1).

3 Technology Trends in Small Satellite

This section covers [5] the trend in technological developments for the small satellites and their ever-increasing applications. They are categorized into different areas of bus systems and payloads.

Bus Systems:

- *Electrical Power*: Generally, 30% of satellite mass is used up in supplying electrical power to all the subsystems. This comprises power generation, storage, and distribution using solar cells and arrays, battery, and power electronics subsystems, respectively.
  - *Solar Cells*: Single-to triple-junction solar cells (with standard and customized sizes) are currently used in small satellites. They have the efficiencies around

15–35%. In the coming years, multi-junction solar cells with more than 5–6 junctions with efficiencies increasing to around 45–70% are expended to make break through and come out competitively priced. Flexible and thin-film solar cells are also making a mark in the development of efficient solutions for tomorrow.
  - *Batteries*: The Li-ion and Li-po batteries are the secondary type batteries commonly used in small satellite missions. They have the advantage of being designed for proven aerospace and terrestrial applications. Thermo-photovoltaic battery is another area of development in progress.
  - *Power Electronics*: The advent of latest power electronics semiconductor devices (GaN FETs, point of load converters, regulators) is bringing in significant changes in the power electronics subsystems in distribution and regulation of power within the small satellites.

- *Thermal Control*: This refers to all the subsystems and technologies pertaining to the maintaining of onboard temperatures to ensure satisfactory functioning of electronic subsystems. These controllers are passive and active in nature. The use of multi-layer-insulation (MLI), optical solar reflectors (OSR) and thermal Coating which are in the form of blankets, tapes, and paints work as thermal diodes in the stoppage of heat flow from outside to inside of the satellite and enable the efficient heat flow from inside to outside. Sunshields (offer shading of satellite from solar radiation), thermal Straps (flexible strips of metal foils or fibers), thermal Louvers (blinds that can be raised or lowered), and deployable radiators are at various technological levels on their way to getting deployed in tomorrow's small satellites. Use of fluid loops (mechanical pumping mechanisms to circulate fluid) and are technological advances in heat pipes are in the queue for qualification and commissioning onto small satellites.
- *AOCS*: Attitude and orbit control systems covers sensors, actuators, and control electronics subsystems. The sensors that help the satellite control are sun sensors, earth sensor, and star sensors. Actuators consist of magnetic troquers, wheels, and gyroscopes.

  - The star trackers use advanced miniature technologies, and several models are available (25–75 arc seconds accuracy ranges) for use onboard the small satellites.
  - The light-weight coarse and fine sun sensors weighing 15–50 g can give accuracy ranges of 0.1–5°.
  - Earth sensors with advanced thermopile sensors weighing less than 50 g are providing 0.25–0.75° accuracies.
  - Actuators such as magnetometers used for measuring local magnetic field are developed with accuracy ranges of 0.65–15 nT and weigh about 200 gms.
  - Gyroscopes are based on MEMS and Fibre Optic technologies. MEMS gyroscopes use vibrating structures that estimate the rotation rate. They weigh around 16gms to 2.8 Kgs for different performance specifications (10–35 deg/hr). The FOG (Fibre Optic Gyroscopes) perform better and weigh around 750gms with bias stability of 1 deg/hr.

- *Propulsion Systems*: The propulsion system comprises thrusters, power processing systems, propellants, storage systems, and feed systems. Due to toxic nature of hydrazie, the hydrazine propellants are being changed to non-toxic propellants with potential thrust levels of 0.2–25 N and specific impulses of around 250 s. Cold and warm gas-based and solid fuel-based propellants are under various stages of development. Electric propulsion is also gaining the developer's attention for their applicability to small satellites. In the coming years, latest versions of the electrosprays, RF ion thrusters, pulsed plasma and vacuum arc thrusters, and hall-effect-based thrusters are being watched for applicability to small satellites.
- *Data Handling and Autonomous systems*: The digital hardware and associated software are going through transformation in subsystems required for satellite mission. Onboard computing is provided using microcontrollers, Field Programmable Gate Arrays (FPGA) and open source platforms. High performance multi processor architecture allow more onboard processing resulting in use of lower bandwidth for downlinking, thus reducing the burden on the communication systems.
- *Communication systems*: Since most of the small satellites are in LEO, the whip, tape, or patch antennae are in common use. Ease of deployability and low directionality make these as popular choices. The UHF-*S* band, *L* and *X* band communication systems with multiple technology readiness are considered for small satellite systems. Laser communication and inter-satellite communication are other areas where significant research is in progress.
- *Mechanical Systems:* Mechanical system design maximizes the electrical capability to accommodate various payloads and to integrate and miniaturize parts and subsystem functions/performance. The standard platform designs are based on an optimized and lightweight modular design with reducing the cost and schedule of the entire satellite development.

Payload Systems:

Payloads for Small satellites are generally classified into Earth Observation and Communication [5].

- High-resolution optical imaging payload form is the most important part of the remote sensing or earth observation payload. The approaches such as deploying space assets at lower altitudes, deployable lenses, and post processing software. This can help achieve an optical imagery around 0.5 m. Post processing has so far been able to achieve 0.9 m imagery.
- Synthetic Aperture Radar payloads need significant power and size. Use of Frequency Modulated Continuous Wave (FMCW) than using traditional pulse-based radar technology are increasingly popular to manage the mission with lesser power.
- Higher level of onboard processing with techniques such as data compression, data synthesis in orbit, and modulation schemes reduces bandwidth and spectrum requirements.

- Advances in optical communication, particularly for intersatellite links ease spectral bandwidth constraints.
- Miniaturization using multiple technologies such as micro-nano electronics and advanced packaging techniques are significantly bringing down the size and mass of the payload electronics.

Assembly Integration and Testing:

- The time from concept to commissioning of small satellites in orbit has been shrinking significantly. The timeline is reducing from about a decade in traditional satellite to anywehere between 6 months and 5 years depending on the constellation complexity for the Small satellites.
- The number of small satellites in the constellation is also necessitating changes in the assembly, integration, and testing activities. Automation of the manufacturing line like the high-end car manufacturing is implemented resulting in a lower cost higher quantity production.
- Modular designs using standard bricks often is helping cost reduction and increased speed of small satellite manufacturing. Future satellite factories comprise robot-filled assembly lines like automotive industry.
- Customized environmental testing for tests like thermo-vacuum and vibration is carried out to different levels during design validation and qualification stage of the subsystems.

Constellations:

- Every satellite mission has its own objectives and characteristics. One satellite may be some time enough to achieve its objectives. However, with small satellites and each having its own capacity constraint, there will be certain mission objectives or services needing coordinated way to have group of satellites to offer certain global-level services.
- In LEO, a satellite completes around 14–15 polar orbits per day, but not over the same area. Hence, each step is different from the previous one. For earth observing missions, involving slow and gradually changing scenario like crops, snow melt, etc., and seven-day repeat cycle is manageable. However, for applications such as traffic control, where frequent images are necessary, the seven-day waiting period is too long, and hence the mission is not viable with only one satellite.
- During the initial mission design phase, it is important to decide how many satellites a constellation will have. Some of the questions [6] that can be answered to finalize a small satellite constellation mission are:
  - Which global areas will this mission serve?
  - How often the satellites receive and transmit information,
  - How are frequencies coordinated to prevent interference?
  - How soon can the constellation enter the service? How is satellite replacement planned?

4 Small Satellite Applications

The broad category of Small satellite applications [7, 8] are earth observation/remote sensing and communications.

Under the earth observation class, the following applications are fast emerging and advancing.

- *Agriculture:* Ability to predict the crop yields, precision agriculture, assess crop irrigation needs, water use and sustainability using the satellite imagery.
- *Resource Management:* Water management, tracking reserves, drought assessment and also mineral, coal, and oil resource assessment, and tracking illegal logging and burning are excellent applications in economic monitoring area.
- *Automatic Identification System (AIS) tracking*: of ships to identify illegal port activity, liiegal fishing, and oil spills.
- *Weather predictions and disaster monitoring*: Monitoring atmospheric and climate shifts, extreme weather events or earthquakes, and hurricane predictions.

Under the applications relating to meeting consumer demands related to communications, the following applications are becoming popular.

- Broadband Internet—LEO constellations: DTH and Community aggregator.
- Low latency upgrades to compete with GEO based broadband providers.
- IoT for asset tracking, autonomous vehicle connectivity.

5 Small Satellite EcoSystem

Like the conventional space ecosystem, small satellite also has its own ecosystem that comprises different players. New players and startups are considering the advantages of the growing market, technological advancements, and launch cost reductions. The dependance on governmental agencies and set ups is decreasing in the small satellite eco system and larger growths are seen in private sector. Table 1 gives the details of players and their roles.

The level of activity by a player varies across the eco system. Some of the players focus on vertical integration and others on horizontal integration. Figure 3 shows the small satellite ecosystem model.

6 Conclusion

The small satellite ecosystem is an emerging sector that has rapidly grown in the last decade. There are many inter-related drivers that are influencing the direction and growth of the small satellite technology and applications. These have the advantages of enjoying the benefits of both the commercial/industrial and aerospace domains.

Table 1 Small satellite ecosystem and players involved

Industry	Traditional space players Start ups Non-traditional space players	• Traditional space has generally medium to large established players operating • Startups have emerged developing new applications and technologies • New space has attracted new companies traditionally not engaged in space sectors
Academic	Universities and academia	• Universities and academic and research institutes have been using Small satellites to test their concepts, products, and technologies
NonProfit	Non-profit research organisations	• Non-governmental organizations and foundations are active as end users and funding agencies
Government	Space agencies Defense and intelligence agencies	• Governmental organizations use small satellite platforms for technology developments and scientific projects

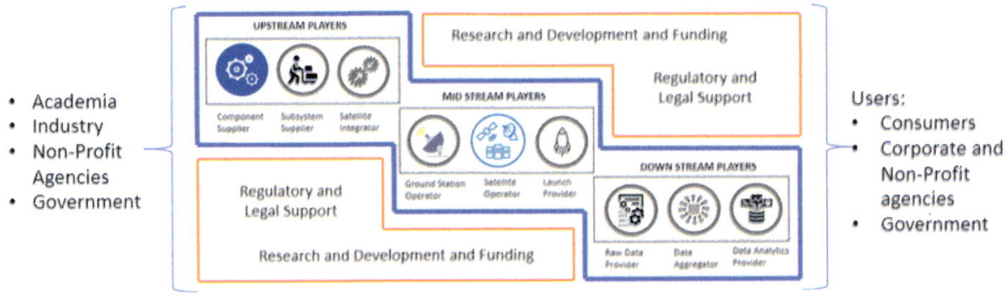

Fig. 3 Small Satellite ecosystem and its players

While the end users continue to evaluate the benefits of small satellite-based applications when compared to the other terrestrial and aerospace domains, the challenges such as debris mitigation, regulatory and legal aspects must be clearly defined. In the future, there will be increased evolution of COTS standards and modular design approaches and blurring of lines between larger small satellite and more compact traditional satellite domain offerings.

References

1. Behrens JR, Lal B (2019) Exploring trends in the global small satellite ecosystem. New Sp 7:126–136. https://doi.org/10.1089/space.2018.0017
2. De Concini A, Jaroslav T (2019) The future of the European space sector, pp 1–162
3. Wekerle T, Filho JBP, da Costa LEVL, Trabasso LG (2017) Status and trends of smallsats and their launch vehicles—an up-to-date review. J Aerosp Technol Manag 9:269–286. https://doi.org/10.5028/jatm.v9i3.853

4. Barnhart DJ, Sweeting MN, Centre SS (2014) Right-sizing small satellites fundamental question: "what is the right size for a small satellite ?" (<200 kg) Three proposed design factors. In: Proceedings of AIAA/USU conference on small satellite
5. Lal B, Balakrishnan A, Picard A, Corbin B, Behrens J, Green E, Myers R, Zuckerman B, Yarymovych M, Boyd I, Macdonald M (2017) Trends in small satellite technology and the role of the NASA small spacecraft technology program final update to the NASA advisory committee technology, innovation and engineering committee
6. Gordon Campbell T (2020) Mega-constellation satellites on the horizon 2030, pp 1–5
7. Pellegrino S, Johnson M, Erwin D, Klesh A, Mueller J, Puig-Suari J, San CP, Obispo L, Llc T, Shaffer C, Bennett M, Martin C, Petro A (2014) Small satellites: a revolution in space science, pp 1–84
8. De E, Blanco R, Behrens JR, Corbin BA, Green EK, Picard AJ (2017) Global trends in small satellites Bhavya Lal

Coherent Population Trapping Based Atomic Sensors for Space Application

Kaitha Rajaiah, Pragya Tiwari, R. Manjula, Nikhil Thakur, Minni J. Kappen, S. B. Umesh, P. Selvaraj, S. Nirmala, T. Venkatappa Rao, S. Pradhan, P. Kalpana Arvind, K. V. Sriram, A. S. Laxmi Prasad, and Prashanth C. Upadhya

Abstract Coherent population trapping (CPT) is a promising technique to realize miniaturized atomic frequency standard (clocks) and magnetic field sensor (magnetometer) for precision measurements in space. Atomic clocks in particular are the reference source for precision time and frequency signals in navigational satellites. Frequency stability of atomic clocks and sensitivity of magnetometer depend primarily on the quality factor of CPT resonance which is the function of amplitude (contrast) and line width ($\Delta \nu_{FWHM}$). In this paper, the CPT resonance characteristics are studied with respect to various critical operating parameters such as laser excitation intensity (atomic excitation), cell temperature, and radio frequency (RF) power. Our study shows that FWHM increases linearly with laser intensity, whereas resonance signal contrast increases with cell temperature up to an optimum temperature and then after decreases with increase in temperature. The numerical values for these operating parameters are then derived by analyzing the experimental results and then arrived at an optimized quality factor for the atomic system.

Keywords Atomic clock · Hyperfine splitting · Coherent population trapping

K. Rajaiah (✉) · S. Nirmala
U.R. Rao Satellite Centre, ISRO, Bangalore, India
e-mail: krajaiah@ursc.gov.in

P. Tiwari · R. Manjula · N. Thakur · M. J. Kappen · S. B. Umesh · P. Selvaraj · P. Kalpana Arvind · K. V. Sriram · A. S. Laxmi Prasad · P. C. Upadhya
Laboratory for Electro-Optics Systems, ISRO, Bangalore, India

T. Venkatappa Rao
Department of Physics, NITW, Warangal, India

S. Pradhan
Bhabha Atomic Research Centre, Mumbai, India

© The Author(s), under exclusive license to Springer Nature Singapore Pte Ltd. 2023
R.S. Priyadarshini and T. Sundararajan (eds.), *Advances in Small Satellite Technologies*, Lecture Notes in Mechanical Engineering, https://doi.org/10.1007/978-981-19-7474-8_23

1 Introduction

Quantum interference can be well explained by considering a three-level energy system in alkali atoms. It can be excited in these atoms when two coherent optical fields couple two long-lived hyperfine ground states ($|1\rangle$ and $|2\rangle$) to a common excited state ($|3\rangle$) as shown in Fig. 1. When Raman resonance condition is satisfied, i.e., the frequency difference of synchronized laser fields ($\omega_1 - \omega_2$) is equal to the splitting of hyperfine ground states ($\nu_1 - \nu_2$) then the atoms do not absorb light. This is due to the destructive interference of transition probabilities, and as a result, the atoms are trapped in a new state called dark state which is decoupled from optical fields. This phenomenon is called as coherent population trapping (CPT) [1].

When laser field interacts with an atom, the corresponding Hamiltonian can be represented as [2]:

$$\hat{H}_{\text{int}} = \frac{\hbar\Omega_{R1}}{2} e^{-i(\omega_1 t+\varphi_1)}|3\rangle\langle 1| + \frac{\hbar\Omega_{R2}}{2} e^{-i(\omega_2 t+\varphi_2)}|3\rangle\langle 2| + \text{h.c} \tag{1}$$

where ω_1, ω_2 and φ_1, φ_2 are angular frequencies and phases of laser fields acting on $|1\rangle \rightarrow |3\rangle$ and $|2\rangle \rightarrow |3\rangle$, respectively. Ω_{R1} and Ω_{R2} are the Rabi frequencies which indicate the strength of the atom-laser field interaction and h.c is Hermitian conjugate. Under CPT condition, the non-coupled state ($|\text{NC}\rangle$) satisfies:

$$\langle 3|\hat{H}_{\text{int}}|\text{NC}\rangle = 0 \tag{2}$$

Thus, there is no transition from $|\text{NC}\rangle$ to $|3\rangle$ which indicates that the atoms in $|\text{NC}\rangle$ cannot absorb photons, and it is called as dark state.

CPT phenomenon in Rb and Cs atoms has been employed in the development of miniaturized high precision atomic clocks [3–5] and atomic magnetometers [6, 7] which can be used as sensors in small satellites. Frequency stability of clocks and sensitivity of magnetometer based on CPT phenomena depend primarily on the characteristics of CPT resonance signal such as its amplitude (contrast) and full width at half maximum (FWHM), i.e., line width ($\Delta\nu_{\text{FWHM}}$). These characteristics are studied with respect to quality factor (q) which is the ratio of contrast (C) to

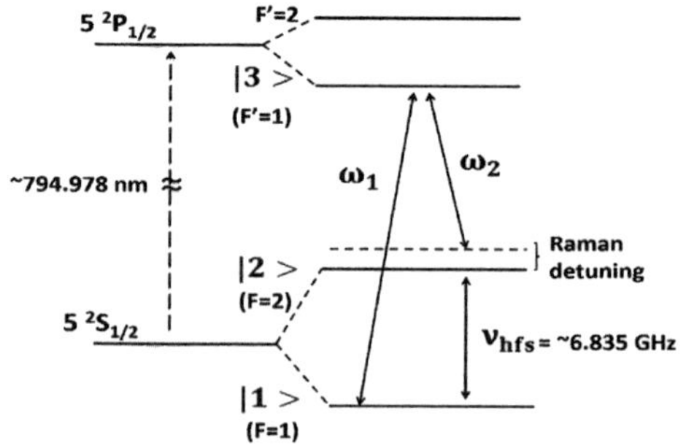

Fig. 1 Hyperfine energy levels in Rb-87 atoms forming a Λ-system for atom-laser field interaction

FWHM of CPT resonance:

$$q = \frac{C}{\Delta \nu_{\mathrm{FWHM}}} \tag{3}$$

A CPT resonance signal with maximum contrast and minimum full width at half maximum would result in higher the value of q, leading to superior performance of atomic systems. However, signal contrast and width depend on the number of atoms coupled to the dark state and the life time of ground state coherence, respectively [8]. Thus, the quality factor depends on various critical operating parameters such as laser excitation intensity, cell temperature, and RF power. This paper describes the influence of these operating parameters on quality factor and its optimization.

2 Experimental Setup

The functional block diagram of CPT implementation scheme employed in this study is as depicted in Fig. 2a. VCSEL diode emitting light at 794 nm is used to produce two coherent optical fields by means of current modulation such that the frequency difference between these two coherent optical fields is equal to splitting of hyperfine ground states. The laser light with beam diameter 3 mm first passes through a neutral density (ND) filter in order to change the laser excitation intensity to a desired value. The attenuated laser light is then travels through a quarter-wave (λ/4) plate which converts the polarization of light from linear to circular for state selection.

The circularly polarized light is passed through a pyrex vapor cell with a length of 25 mm and diameter of 25 mm containing natural Rubidium (Rb-85 + Rb-87) and Ne as buffer gas at 50 Torr pressure. The transmitted light through vapor cell is detected by a Silicon photo detector. Magnetic field is applied along the laser beam direction using a solenoid coil for exciting CPT among field independent *Zeeman*

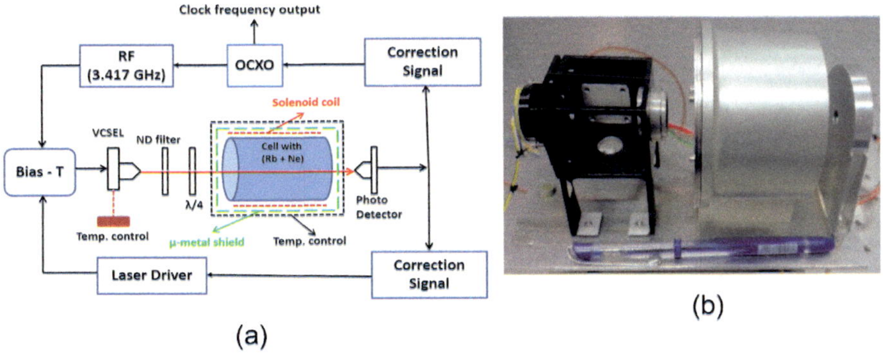

Fig. 2 **a** Block diagram of CPT atomic clock. **b** Engineering model of sensor head (~800 g)

lines ($m_F = 0$). When the laser field with RF modulation is scanned such that two-ground state hyperfine transitions of Rb-87 atoms are resonant with the first order side bands, CPT resonance is observed [9]. The characteristics of CPT resonance are studied by varying the cell temperature from 65 °C to 83 °C. Laser frequency and the OCXO are stabilized by employing phase sensitive detection (PSD) and servo locking system as shown Fig. 2a. The engineering model of physics package (sensor head) is as shown in Fig. 2b.

3 Laser Intensity and Cell Temperature

CPT resonances were measured at different laser excitation intensities, ranging from 4 W/m^2 to 19 W/m^2. Figure 3a shows CPT resonances for different laser intensities while maintaining the temperature of atomic cell at 71 °C. It can be observed that the width of CPT resonance increases with increase in laser intensity. FWHM is calculated by fitting Lorentzian function to the CPT resonance. Figure 3b shows the FWHM of CPT resonance and quality factor (q) as function of laser intensity. Maximum quality factor is observed at laser intensity of 4.25 W/m^2.

Figure 4a shows the variation of quality factor with cell temperature, measured at a fixed laser excitation intensity of 7.08 W/m^2. It is clear that the quality factor increases with cell temperature initially, and reaches a maximum value at a particular temperature beyond which it decreases sharply. The optimum operating temperature of the atomic cell is defined as the temperature at which its measured quality factor is maximum (as shown in Fig. 4a) while the laser excitation intensity is kept constant. The atomic density increases exponentially with increase in the temperature of atomic vapor cell. Hence, more atoms are available for interaction with laser fields which results in higher contrast [10]. As the temperature reaches above certain value, higher atomic density induces more single photon absorption between the hyperfine levels

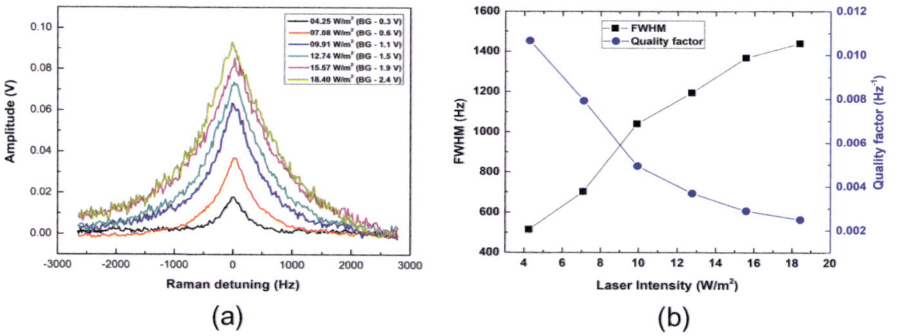

Fig. 3 **a** CPT resonances for different laser intensities at cell temperature = 71 °C (Back ground, BG, removed from each resonance for comparison). **b** FWHM and quality factor (q) for different laser intensities at cell temperature = 71 °C

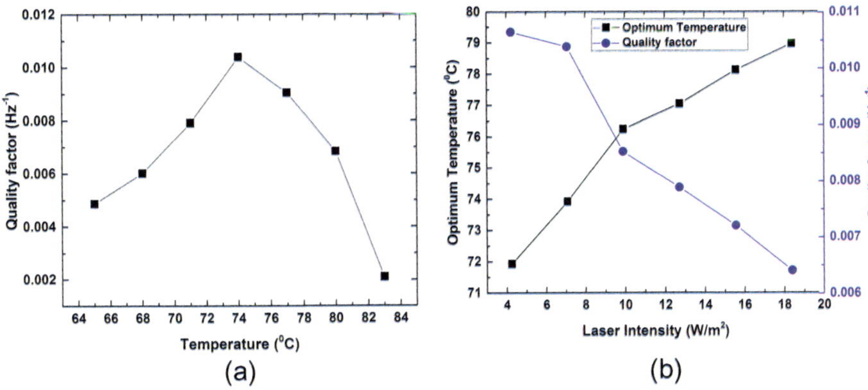

Fig. 4 **a** Quality factor as a function of temperature at laser intensity of 7.08 W/m^2 and **b** Optimum operating temperature and its corresponding magnitude of quality factor as function of laser intensity

thereby quenching the CPT resonance which is two-photon resonance transition leading to reduction in contrast.

From experiments, it is also observed that the optimum operating temperature of the atomic cell varies with laser excitation intensities. The optimum operating temperature values are obtained for each laser intensities individually by fitting Gaussian to the corresponding quality factor versus temperature curve. Figure 4b depicts the variation of optimum operating temperature and corresponding magnitude of quality factor as function of laser excitation intensity. Highest quality factor (q = 0.01065 Hz^{-1}) is observed at optimum temperature of 72 °C and laser excitation intensity value of 4.25 W/m^2, i.e., at the lowest laser excitation intensity. However, at intensities below 4.25 W/m^2 the resonance amplitude drops to noise floor owing to the reduced laser-atom interaction, making the detection difficult.

4 RF Power

The RF power that couples to laser diode also influences the resonance signal characteristics. Figure 5 depicts dependence of the quality factor on RF power. The q value increases for lower RF power and then decreases sharply with RF coupling power. Here, the q value is predominantly governed by resonance amplitude which decreases at higher RF power. Maximum q value is observed at −8.465 dBm for optimum cell temperature of 71 °C.

Finally, experimentally derived values of optimized cell temperature (72 °C), laser intensity (4.25 W/m^2), and RF coupling power (−8.465 dBm) would result in an optimal quality factor, q = 0.01065 Hz^{-1}. Under these operating conditions, the estimated CPT clock's short-term stability is expected to be <10^{-12} and field sensitivity of magnetometer <100 nano Tesla.

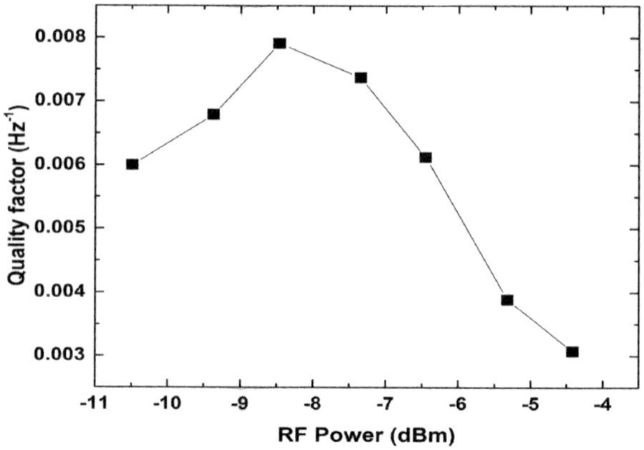

Fig. 5 Quality factor of CPT resonance as function of RF power

5 Conclusion

The performance efficiency of CPT-based miniaturized atomic clock and atomic magnetometer primarily depends on the CPT resonance characteristics (quality factor) such as contrast and FWHM. The experimental investigation is carried out on optimizing the operating parameters, viz., laser intensity, cell temperature, and RF power which influence the resonance characteristics. Our results indicate that the quality factor exhibits an optimum operating temperature whose magnitude decreases with increase in laser excitation intensity. Finally, an optimized quality factor, $q = 0.01065\ \text{Hz}^{-1}$ is achieved in our CPT atomic system by setting the operating parameters to their optimal values that are derived from the experimental results. This study is useful in the construction of portable, reliable, stable, and sensitive atomic systems needed in small satellites.

Acknowledgements The authors would like to thank Mr. Madan Mohan Mehra, Mr. Bijoy Raha, Mr. Madan Mohan Kandpal and Dr M. V. Hanumantha Rao (all at LEOS) for their valuable inputs and suggestions. K. Rajaiah would like to acknowledge Mr. Kartik A. from Space Navigation Group at U.R. Rao Satellite Centre, Bangalore for his support and encouragement.

References

1. Arimondo E (1996) Relaxation processes in coherent-population trapping. Phys Rev Lett 54:2216–2223
2. Zhu M (2003) High contrast signal in a coherent population trapping based atomic frequency standard application. In: Proceedings of the 2003 IEEE, IFCS and PDA Exhibition Jointly with the 17th European frequency and time forum
3. Merimaa M, Lindvall T, Tittonen I, Ikonen E (2003) All-optical atomic clock based on coherent population trapping in ^{85}Rb. J Opt Soc Am B 20:273

4. Kitching J, Knappe S, Vukičević N, Hollberg L, Wynands R, Weidemann W (2000) A microwave frequency reference based on VCSEL-driven dark line resonances in Cs vapor. IEEE Trans Instrum Meas 49:1313–1317
5. Vanier J (2005) Atomic clocks based on coherent population trapping: a review. Appl Phys B Lasers Opt 81:421–442
6. Nagel A, Graf L, Naumov A, Mariotti E, Biancalana V, Meschede D (1998) Experimental realization of coherent dark-state magnetometers. Europhys Lett 44(1):31–36
7. Belfi J, Bevilacqua G, Biancalana V, Dancheva Y, Moi L (2007) All-optical sensor for automated magnetometry based on coherent population trapping. J Opt Soc Am B 24:1482–1489
8. Knappe S, Wynands R, Kitching J, Robinson HG, Hollberg L (2001) Characterization of coherent population-trapping resonances as atomic frequency references. J Opt Soc Am B 18(11):1545–1553
9. Pradhan S, Kani A, Wanare H, Mishra S, Das AK (2015) Magic frequency enabled by quantum interference for a dual atomic device. J Phys B At Mol Opt Phys 48:075502
10. Knappe S, Kitching J, Hollberg L, Wynands R (2002) Temperature dependence of coherent population trapping resonances. Appl Phys B 74:217–222

GNSS Receiver Autonomous Integrity Monitoring

Archa P. Lal, V. S. Vinoj, and R. Harikumar

Abstract Global navigation satellite system has become a de-facto standard for navigation, with enormous applications worldwide. Currently, four global navigation satellite systems are operational and a number of regional navigation satellite systems are also in place including Indian Regional Navigation Satellite System (IRNSS) which is also known as Navigation with Indian Constellation (NavIC). NavIC is a regional navigation satellite system owned and operated by ISRO. The application of GNSS ranges from commercial applications to intercontinental ballistic missiles or heavy lifters of ISRO/NASA. High dynamic NavIC receivers are used in ISRO launch vehicles for preliminary orbit determination/closed loop guidance along with inertial navigation system in an integrated manner. When such critical applications of GNSS are considered, the integrity of navigation data in real time is a prime concern. Different methods are developed by different agencies over time in this area. Usage of inbuilt health information from satellite is one of the possible methods of integrity check. Receiver autonomous integrity monitoring (RAIM) provides integrity checking of GNSS without an external aid. NavIC/GPS receiver architecture is presented in this paper, the satellite signal analysis is done, and possible integrity threats are identified. The different approaches are considered in RAIM algorithm, and a simplified RAIM approach is developed which has good real-time performance. The simulation studies and flight software development are completed.

Keywords Integrity monitoring · RAIM · Pseudo-range · GNSS · Fault detection and isolation · NavIC · GPS · Satellite navigation

A. P. Lal (✉) · R. Harikumar
Electrical and Electronics Engineering, College of Engineering Trivandrum, Trivandrum, Kerala, India
e-mail: archaplal@gmail.com

V. S. Vinoj
Satellite Navigation Division, IISU, ISRO, Trivandrum, Kerala, India

© The Author(s), under exclusive license to Springer Nature Singapore Pte Ltd. 2023
R.S. Priyadarshini and T. Sundararajan (eds.), *Advances in Small Satellite Technologies*, Lecture Notes in Mechanical Engineering, https://doi.org/10.1007/978-981-19-7474-8_24

1 Introduction

GNSS receivers are used in all phases of a modern civil society now. However, commercial GNSS receivers are meant for terrestrial use and are bound by ITAR restrictions and are invariably designed for low dynamics applications—typically accelerations less than 3 g and jerks less than 2 g/s. For high dynamic applications, such as launch vehicles, commercial GNSS receivers are unsuitable and hence, specially designed high dynamic GNSS receivers are employed. Most important characteristics of such receivers are the accuracy and integrity with hard-real-time performance. The accuracy of the system is ensured by measurement accuracy and navigation algorithms. The hard-real-time nature is mainly decided by the high dynamic tracking loop. The integrity of the system is currently ensured in the receiver using the inbuilt health bit of the satellite. However, for mission/safety critical applications like human space mission, an independent real-time autonomous integrity monitoring is essential. A simplified, processing efficient approach in autonomous integrity monitoring for NavIC and GPS is developed and is detailed in this paper. Section 2 consists of the hardware design of a GPS/NavIC receiver. Section 3 contains the software architecture. Section 4 gives the requirement of RAIM. Section 5 gives the methods developed for autonomous integrity monitoring of receiver. Section 5.3 gives the detailed algorithm, simulation results. Conclusion is given in Sect. 6.

2 GNSS Receiver Hardware Architecture

The basic architecture of any GNSS receiver consists of RF processing subsection, correlation subsection and a computing subsection. The RF processing is universally done in hardware, and the computation subsection is usually implemented in a DSP or a microprocessor. The correlation subsection has the flexibility of being implemented in hardware or software. The receiver used for developing the RAIM algorithm is a hard correlator-based NavIC receiver, enabled with GPS and GAGAN.

The overview of the receiver architecture is given in Fig. 1. In the receiver, the RF processing subsection is centered on an RF front end ASIC. The correlation subsection is implemented in FPGA. The computation subsection is centered on a DSP from Texas Instruments. Part of the computation that includes Viterbi decoding for NavIC and GAGAN channels are implemented in FPGA.

3 GNSS Receiver Software Architecture

The general architecture of the GPS/NavIC receiver is given in Fig. 2. The incoming signal is searched for expected satellite signal and is called signal acquisition. The continuous tracking of signal is achieved using code tracking and carrier tracking

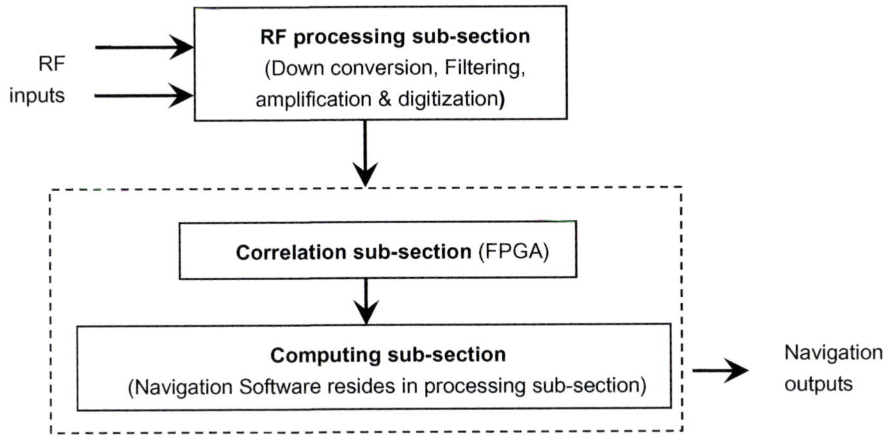

Fig. 1 General hardware architecture of NavIC/GPS receiver

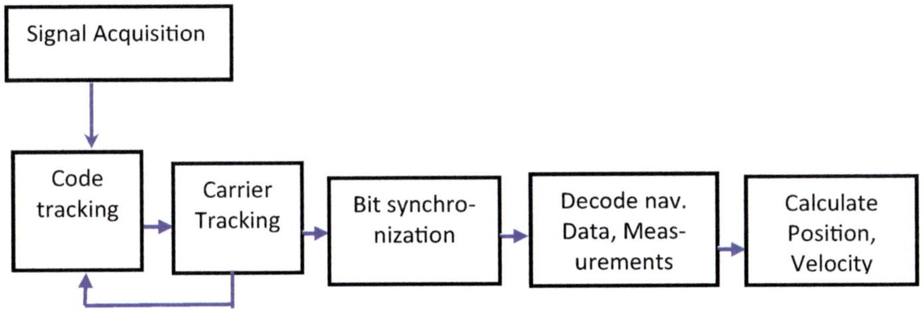

Fig. 2 General software architecture of NavIC/GPS receiver

blocks. Bit synchronization is achieved in 20 ms periodicity on correlator accumulator value over 1 ms. Navigation data decoding and position velocity is calculated using the measurements. The RAIM algorithm is implemented on the last block of software where the position, velocity is computed.

4 Requirement of Receiver Autonomous Integrity Monitoring

In GPS system, each satellite transmits its own health in 6 bits, called satellite vehicle health (SV health), which is used to validate the satellite measurement as well as ephemeris. Another 8-bit health is also available in GPS sub frame 4, which indicates the health of almanac data. Another factor of health is the user range accuracy, which is the statistical indicator of the range accuracy that can be achieved if the particular satellite is used for navigation. All the three data are used in the existing algorithm, but all these are set by the satellite based on some ground-based systems. Hence, the real-time performance of the error flag/URA is very poor. In case of any clock jump in the satellite, the satellite may set the error after a fixed time which can go even in minutes, which is not acceptable for a launch vehicle navigation system. A typical error occurs in a receiver after a satellite clock jump of 1 μs can lead to an error of 300 m in position data. So, if such an error happens the receiver should be able to identify the error and isolate the satellite from navigation computation using the measurement itself without depending on health flag/URA.

5 RAIM Algorithm Design and Tradeoff Studies

5.1 Code Phase/Pseudo-Range Measurements

Pseudo-range is the measured range of the user receiver from the satellite and is obtained by taking the product of the speed of light with code phase measurement. This measurement is not accurate due to known uncertainties like receiver and satellite clock bias, tropospheric and ionospheric error, antenna phase center variations and multipath mitigations. To obtain positioning information of GNSS, time of arrival (TOA) method is usually used and the pseudo-range observation equation can be defined as [4]

$$\rho = R_i + C \times (t_r - t_s) + e_{\text{ion}} + e_{\text{trop}} + e_{rr} \tag{1}$$

ρ is the measured pseudo-range, R_i is the geometrical range which is the true distance between satellite and receiver, t_r is clock offset of the receiver, t_s is clock offset of the satellite, e_{ion} denotes the error of ionospheric delay, e_{trop} denotes the error of tropospheric delay error, e_{rr} is pseudo-range observation error. The error due to troposphere, ionosphere and pseudo-range observation can be eliminated easily. For a launch vehicle single frequency receiver, the troposphere and ionosphere errors can be modeled and compensated. Thus, we can refine the equation for pseudo-range observation as

$$\rho = R_i + C \times (t_r - t_s) \tag{2}$$

This measured pseudo-range using the code phase measurement is used for the RAIM algorithm.

5.2 *Method of Least Square*

The measurement equation of NavIC/GPS system can be represented as

$$M = HX + e \tag{3}$$

where M is the receiver position coordinates in ECEF frame of reference; X is the pseudo-range; H is the geometry matrix which is used to transfer the pseudo-range to position solution, which can be found out by iterative methods like Newton Raphson. Once the position is estimated, the pseudo ranges can be back estimated using the pseudo LSE [1]. That is

$$\hat{X} = \left(H^T H\right)^{-1} H^T M \tag{4}$$

The back estimated pseudo-range can be compared against the measured pseudo-range, and the difference should be within a predefined threshold. This is a standard method for failure detection and isolation in inertial sensor systems also. The method is implemented in software and analyzed the result using the satellite errors induced using GNSS RF simulator (Spirent make GSS 9000 series). The result is analyzed and found that the algorithm gives poor performance at moderate error values. Further, analysis shows that the problem with H matrix computation is the root cause of the issue and when the H matrix is computed using iterative process, it is trying to fix the position which is best agreeing by the pseudo-range. This method is computationally efficient and uses less memory only compacted with other processes. But the error is not equally distributing [2]. So this method cannot be used. The simulation results show that when the satellites are given errors, most of the times, it is identified correctly, but in some cases, false isolation also happens. The graph below gives two different conditions when sixth satellite is provided with a high and low fault values. In first case (Fig. 3), satellite 6 is provided with 10000 kms error, then the residue obtained was higher. So the faulty satellite 6 can be easily isolated. But when the error is low (100 kms), which is depicted in the second case (Fig. 4), then the residue obtained from all the satellites cannot be distinguished. Thus, the faulty satellite cannot be isolated.

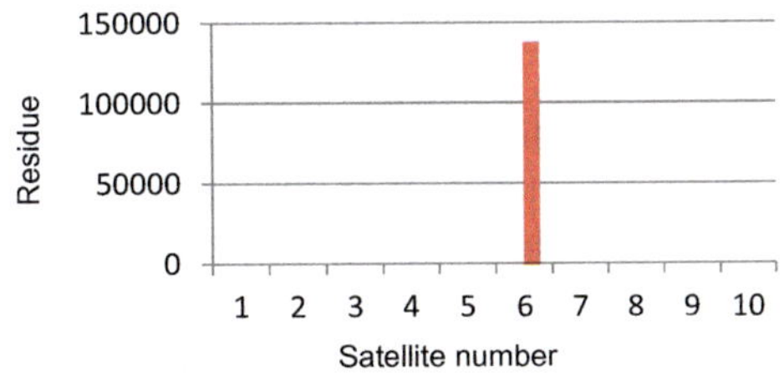

Fig. 3 First case

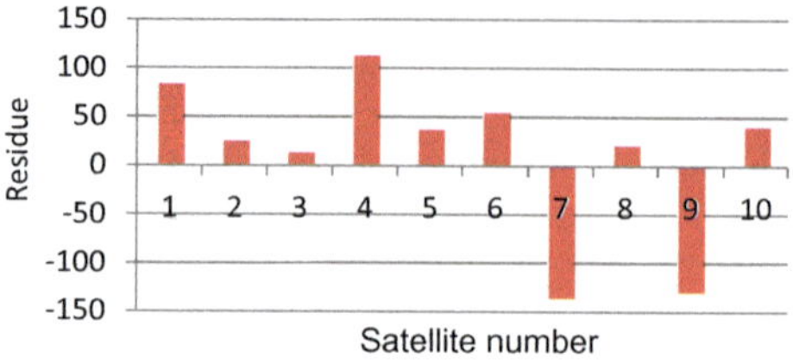

Fig. 4 Second case

5.3 *INS Aided Integrity Monitoring*

In launch vehicles, the primary navigation system is inertial based and GNSS is normally used in integrated way with INS. If INS position is available, then pseudo-range can be estimated using this INS solution, by knowing the satellite position. The measured pseudo-range can be compared against the estimated pseudo-range and if it is beyond a threshold that satellite can be isolated. The advantage of this method is that it is processing and memory efficient, but it is against the autonomous nature of GNSS.

5.4 *Detailed Algorithm and Simulation Results of Proposed Method: Method of Multiple Solutions*

In the third possible algorithm, let a number of sets of satellite combinations were introduced using available pseudo-range. Each group should contain more than 5 satellites. For each set, we calculate the solution. If the positions computed by all combinations are compared each other, then the odd one out can be identified and the satellites involved in that alone will be healthy [5].

One of the major concerns while choosing the minimum number of satellites for each set of solutions is the position dilution of precession (PDOP). Ideally, four satellites are sufficient for position computation, but in most of the time, this will give large PDOP value leading to unacceptably erroneous solutions [3]. So minimum 6 satellites are used for solution computation and PDOP of <6 is expected in all conditions. The detailed algorithm for solution computations is given in Fig. 5. The difficulty with this method is that it is computationally heavy. For single satellite isolation with 19 channels (12 GPS and 7 NavIC), it needs 19 position computations. For two satellite isolation, it needs 171 set of position computations. The receiver considered is having 12 GPS channels and 7 NavIC channels. So independently, the RAIM is implemented with multiple solution method. So maximum of 19 iterations are sufficient but can identify one failure in NavIC and one failure in GPS. This will reduce the number of position computation requirement from 171 to 19.

The graphs shown are obtained when there are 12 GPS satellites in view with the receiver. A fault of 3000 km is provided to the 10th satellite, thus set 10 (Fig. 7)

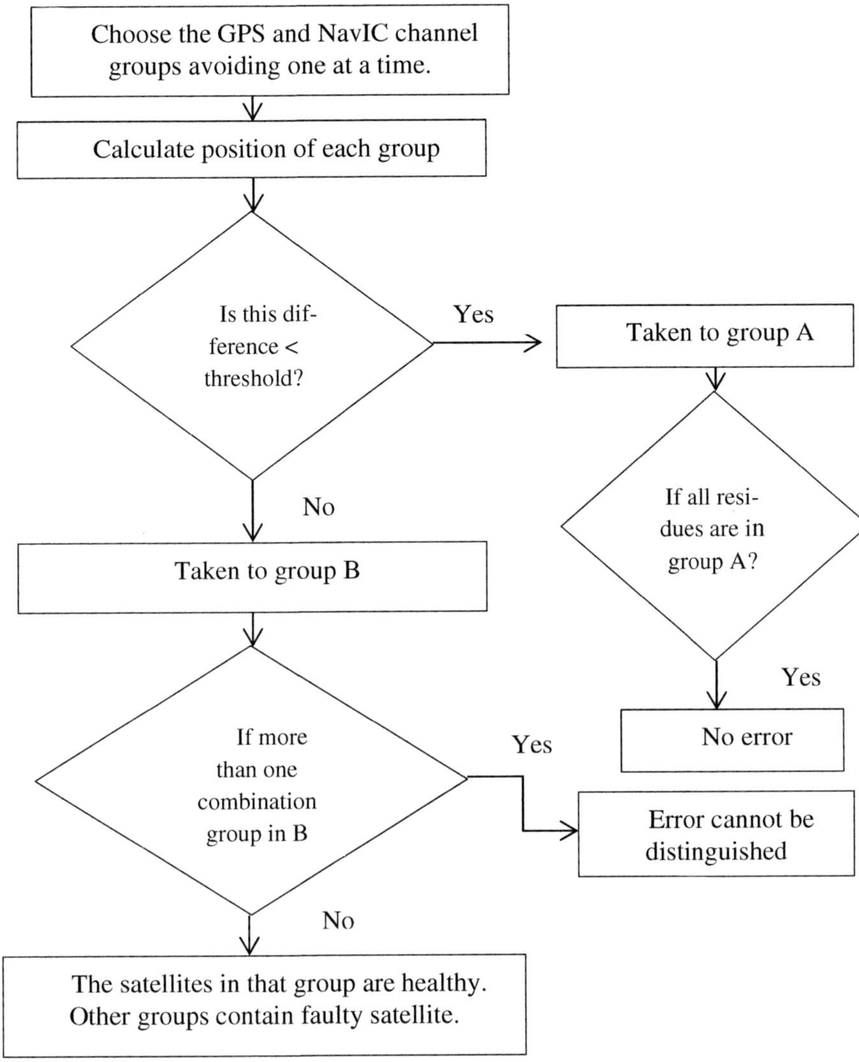

Fig. 5 Flow chart of the proposed method

which eliminates satellite number 10 gives us the most accurate position solution. In set 2 (Fig. 6), the faulty tenth satellite provides error value to all the satellites. The graph of set 10 shows that for each and every satellite, the residue is constant which gives the time bias. Considering the pros and cons of all the three method, the straight forward method using multiple solutions is selected in flight software development with add on algorithm improvement of independent check for GPS and NavIC, thereby getting 2 satellite failure detection capability with 19 set of positions.

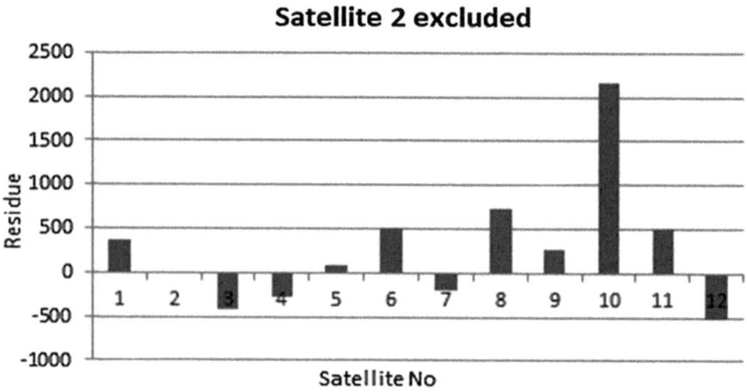

Fig. 6 Set 2

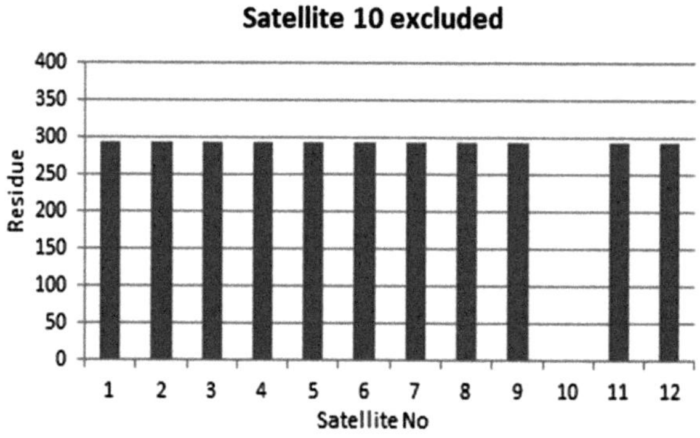

Fig. 7 Set 10

6 Conclusions

GNSS receivers are a commonly used navigational tool to locate a receiver position. Sometimes there may be failed satellites that affect the navigational solution. In this paper, different RAIM algorithms are studied and compared for pros and cons. No ideal algorithm is identified for high dynamic conditions with processing and memory efficiency. All the methods are simulated using GNSS RF simulator (Spirent make GSS 9000 series) and IISU designed NavIC/GPS receiver. Multiple solution technique is identified as the suitable choice. The multiple solution algorithm is redesigned for dual constellation receiver with 12 GPS channels and 7 NavIC channels, and the method can identify up to 2 satellite failures but need only 19 position computations. Simulation tests completed, flight software is developed, and results are satisfactory.

Acknowledgements A lot of effort and hard work has been put into this research work. However, it would not have been possible without the kind support and help of many individuals and other sources. I take this opportunity to express my deep sense of gratitude and sincere thanks to all who helped me to complete this work successfully. I thank Dr. P. Sreejaya, Head of the Department, Department of Electrical. I express my sincere gratitude to my internal guide Prof. R. Harikumar, Department of Electrical Engineering. A special mention to my external guide Mr. Vinoj V S, Scientist, IISU for his supervision, assistance and helpful suggestions throughout this work.

I am deeply indebted to Dr. Sreeja S, Assistant Professor and Dr. Arun Kishore W. C., Associate Professor, Department of Electrical Engineering, for their positive criticism and valuable comments. Finally, I thank my parents and friends near and dear ones who directly and indirectly contributed to the successful completion of my work.

References

1. Brown GR, Parkinson BW, Spilker JJ (1996) Receiver autonomous integrity monitoring. Global Pos Syst Theory Appl 2:143–165
2. Kalafus RM, Chin GY (1988) Performance measures of receiver-autonomous GPS integrity monitoring. In: Proceedings of the 1988 national technical meeting of the institute of navigation, January, pp 223–229
3. Lee Y, Van Dyke K, Decleene B, Studenny J, Beckmann M (1996) Summary of RTCA SC-159 GPS integrity working group activities. Navigation, 43(3):307–338
4. Lee YC (1986) Analysis of range and position comparison methods as a means to provide GPS integrity in the user receiver. In: Proceedings of the 42nd annual meeting of the institute of navigation, June, pp 1–4
5. Parkinson BW, Axelrad P (1988) Autonomous GPS integrity monitoring using the pseudorange residual. Navigation 35(2):255–274

Response of Single Bumper Whipple Shields to Debris Impact

Agesh Markose, Tinto Thomas, and Hannah Bibu Mathew

Abstract Protective shields are indispensable for protecting the satellites components from debris impact. Proper evaluation of the response of these shields is important. High velocity impact simulations were conducted to ascertain the response of single bumper Whipple shields used as satellite shields. The study was conducted with the Abaqus explicit solver with the smooth particle hydrodynamics (SPH) framework. A simple damage-based failure model was able to capture the fragmentation as well as the penetration of the shield under the impact. The energy as well as the residual velocity were seen to match with the similar experiments.

Keywords Debris impact · SPH framework · Energy of the particles

1 Introduction

The significant increase in the space programs has resulted in huge amounts of space debris. These man made debris along with natural debris which consist of pieces of cometary and asteroidal material known as meteoroids pose a serious threat to the small satellites and other space crafts. A high impact collision from the debris often results in significant damage to the satellites. Larger space debris can be tracked down using radio frequency radars, and thus, collision can be avoided. But this becomes difficult in the case of small debris, lesser than 10 cm in diameter [6]. Hence, a better way to reduce the impact from these collisions is the proper application of Whipple shields. Since the average speed of collision with these debris is over 10 km/s, [1] materials with higher strength and toughness are preferred. However, installing shields increases the total weight of the spacecraft. So, material which has higher strength to weight ratio has to be considered and aluminum can fulfill the requirements. The experimental works with very high velocity projectiles on target plates often require substantial infrastructure and equipment cost, even for scaled down models. An alternative would be to depend on numerical simulations

A. Markose (✉) · T. Thomas · H. B. Mathew
Department of Mechanical Engineering, TKM College of Engineering, Kollam 691005, India
e-mail: agm@tkmce.ac.in

© The Author(s), under exclusive license to Springer Nature Singapore Pte Ltd. 2023

R.S. Priyadarshini and T. Sundararajan (eds.), *Advances in Small Satellite Technologies*, Lecture Notes in Mechanical Engineering, https://doi.org/10.1007/978-981-19-7474-8_25

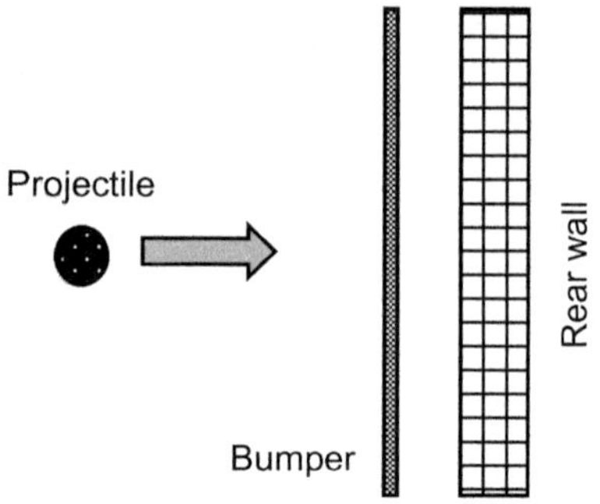

Fig. 1 Schematic showing the configuration of the single bumper Whipple shield

for studying the response parameters. Numerical work has the additional advantage that the variables which cannot be tracked in an actual experiment can be found with the proper selection of numerical models. The aim of this paper is to develop a finite element (FE) model which could be utilized for performing feasibility studies for the design of satellite shields. The FE-based model has been based on the ABAQUS explicit solver available in-house.

2 Methodology

Protection of small satellites from impact of the space debris is presented in the paper. Protective plates with several configurations and materials are available for the purpose [2]. The wall of the space craft is usually protected with plates also known as bumpers as shown in Fig. 1. These bumpers usually absorb the energy form the debris during the impact and debris breaks up in many fragments as a result. The debris may penetrate through the shield but the momentum of these fragments may not be sufficient to cause any major damage to the craft. There could be different parameters which affect the performance of the bumpers like its thickness and its location with respect to the rear wall.

The performance of a bumper shield made with aluminum 6061 panels for protection from cylindrical debris with a 5 mm diameter is explored in this work. A numerical model has been developed and validated with the results available in the literature. The model is further used to predict the parameters like perforation hole diameter and dispersion angle as well as the average velocity of the debris cloud.

3 Numerical Analysis

The dimensions of the debris and shield-plate are given in Table 1. A cylinder made of aluminum is assumed to represent debris since it is the major constituent of the debris. It was reported that hemispherical impactor could be the least effective in

Table 1 Geometric dimensions

	Material	Dimensions
Debris	Al 6061	5 mm in diameter and height 5 mm
Bumper plate	Al 6061	5 × 5 × 0.08 cm

perforation of aluminum targets [5], still many studies on debris impact is seen to employ spherical projectiles. In this study, the focus is on the performance of debris with blunt face. It requires lower perforation energy and hence imposes a more severe loading condition. The assumed velocity of the debris is 1 km/sec although the model is capable of dealing with higher velocities. The preprocessor in ABAQUS is used to construct geometry and meshed model of the debris and plate as shown in Fig. 2. The material models are given in Table 1. The excessive deformation is handed with smoothed particle hydrodynamics (SPH) formulation. The plate has been modeled with Al 6061, which has sufficient strength to weight ratio. The presence of ingredients like magnesium and silicon makes it resistant to cracking. The dimensions of the plate are approximately 10 times the diameter of the debris. This is assumed to impose a higher limit on the energy required for the perforation of plate and hence a more resistant configuration as compared to plate with lesser dimensions [5].

The first order solid elements (C3D8R, that is, continuum three-dimensional 8 node reduced integration) which have minimal shear and volume locking effect has been used for modeling the plate and debris. The outer region of the plate is comparatively stiffer and is has been modeled with the solid elements using Lagrangian formulation throughout the simulation. The projectile and the central region of the plate suffers considerable deformation followed by fragmentation and hence analyzed

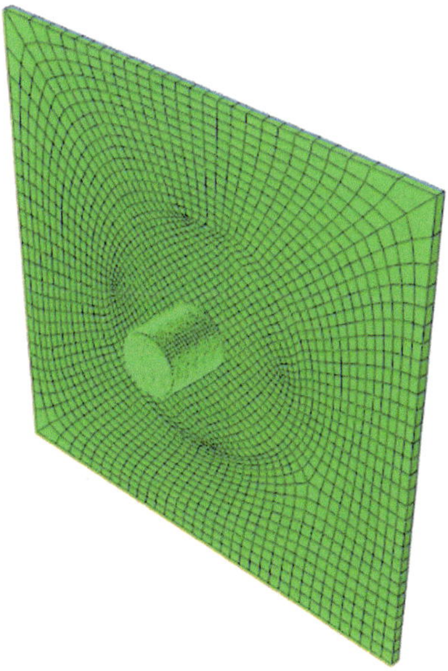

Fig. 2 Meshed finite element model of plate and projectile

Table 2 Material constants

C_0	s	γ_0
5330	1.338	2.18
A	B	n
369 MPa	675 MPa	0.7
d_1	d_2	d_3
−0.77	0.2	−1.3

with PC3D elements. A total of 55,339 elements were used with a hexahedral mesh structure. The high speed impact problems can be modeled with an equation of state (EOS), strength and failure model. The parameters used have been given in Table 2.

The shock response of the plate is represented with the Mie–Gruneisen equation of state, implemented with the U_s–U_p formulation available in ABAQUS, with the parameters [4] as given in Table 2. The dynamic strength model has been modeled with Johnson cook model as given below

$$\sigma = \left[A + B\varepsilon^n\right]\left[1 + Cln\dot{\varepsilon}^*\right]\left[1 - T^{*m}\right] \tag{1}$$

A simplified form of Johnson cook damage model is adopted considering only the constants d_1, d_2 and d_3 [6]. The selected parameters for the strength and failure model are given in Table 2.

4 Results and Discussions

The numerical simulations were conducted with mesh size of 0.0009 m for the regions under maximum deformation. The outer regions were meshed with relatively coarse mesh as shown in Fig. 2. The SPH model captured fragmentation followed by the impact and penetration Fig. 3. Researchers have published calculations based on the experimental observations of the hypervelocity impact of projectile on shields made with aluminum [2]. They have calculated the ratio of kinetic energies (KE) of the debris and projectile and found them close to 0.3. The variation in KE of the current model is plotted in Fig. 4. The initial part of the curve shows the kinetic energy of the debris prior to penetration. It is seen that the kinetic energy of the debris decreases during the initial stages of the collision due to the momentum transfer. But simultaneously, the fragmentation of the debris is initiated and the fragments pick up the kinetic energy. Fragmentation continues up to the point of complete plate rupture. The kinetic energy of the debris particles can be computed from kinetic energy of the whole model after accounting for the KE of the debris after impact. With these data, the ratio of the kinetic energies has been calculated for the present model as approximately 0.4. These are sufficiently close results considering the simple failure criteria and material models used. Similarly, the numerical calculations were also made to find the average velocity of the debris cloud. The output data base contains the fragments with associated velocity fields. The magnitude of the average velocity of the fragments has been calculated for the present model as 194 m/sec.

The impact is seen to produce a complete perforation of the plate as shown in Fig. 5. The average hole diameter (*D*) of the perforated plate has been determined to be 9.84 mm. An attempt was also made to find the average angle of dispersion of the particles. Average angle of dispersion is defined as angle between axial direction of the projectile and averaged direction of motion of the particle-sample under consideration. It was seen from the analysis that the particles which are closest to the periphery of the perforation hole suffers the maximum dispersion. A representative sample of 152 particles which suffers the maximum vertical dispersion were

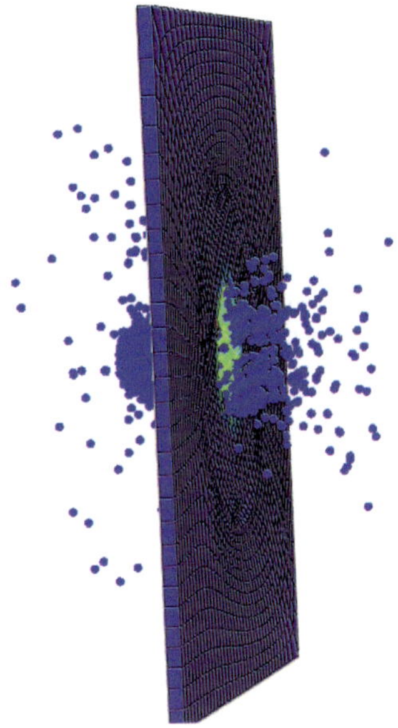

Fig. 3 Impact and penetration of the projectile at time 1.2e-5 s, resulting in fragmentation of the plate and debris

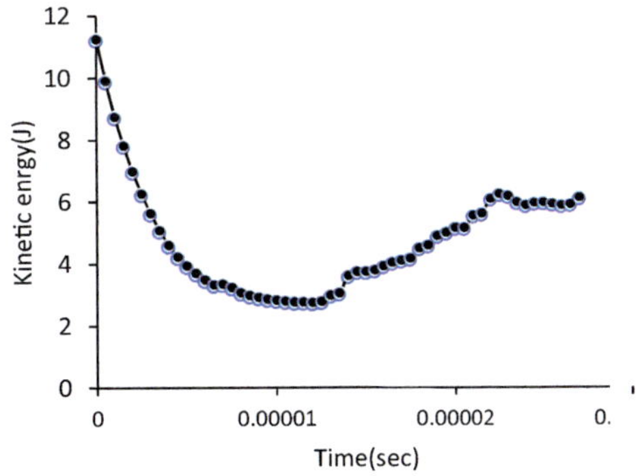

Fig. 4 Energy variation of the whole model during the debris impact

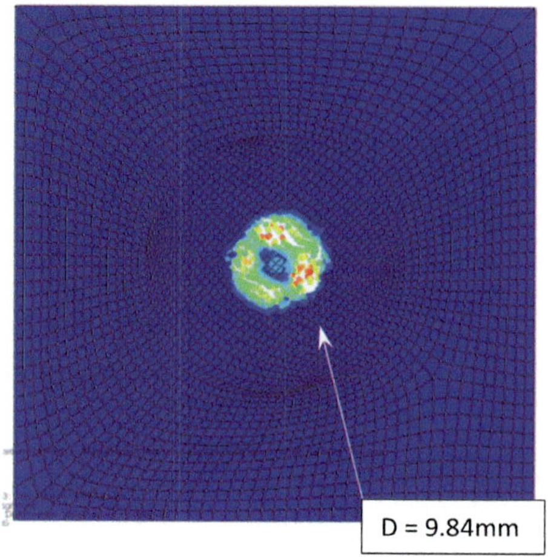

Fig. 5 Image of the perforation in the aluminum sheet

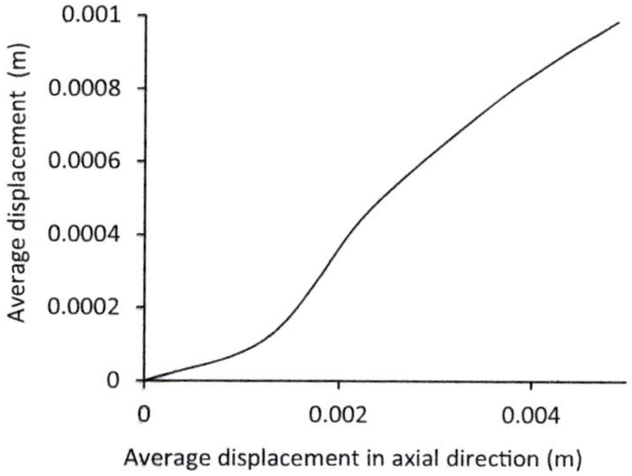

Fig. 6 Average axial displacement and corresponding vertical displacement of a representative sample

analyzed. The average displacement of these particles in the vertical direction and corresponding displacement in the axial direction is plotted in Fig. 6.

5 Conclusions

The simple model utilizing the SPH frame work has been used for modeling the debris impact on single bumper aluminum Whipple shield. The ratio of the kinetic energies of the debris and projectile has been used for validating the numerical simulations. The model has been used to predict the average velocity of the particle cloud. In that case, also the predictions are coming within very close limits of residual velocity due to projectile impacts on aluminum plates. The perforation on the aluminum sheet has been simulated and found to have a diameter of 9.84 mm. An estimate of the dispersion of the debris particle was made based with a representative sample which suffers the maximum dispersion. The calculations show that the simple method can be used for finding the important response parameters of the single bumper Whipple shield against debris impact. However, the method can be improved with better mesh refinement and better sampling methods for the calculation of fragment properties. It is felt that simple numerical models like the one in the current study can be used for getting sufficiently accurate results for comparing the available designs.

References

1. Christiansen EL (2009) Handbook for designing MMOD protection, NASA/TM-2009- 214785, pp 53–58
2. Higashide M, Kusano T, Takayanagi Y, Arai K, Hasegawa S (2015) Comparison of aluminum alloy and CFRP bumpers for space debris protection. Procedia Eng 10:189–196
3. Boldyrev I. Shchurov I, Nikonov AV (2016) Numerical simulation of the Aluminum 6061-T6 cutting and the effect of the constitutive material model and failure criteria on cutting forces' prediction. Procedia Eng 150:866–870
4. Mehra V, Pahari S, Savita AN, Prasad IS, Shiv N, Chaturvedi S, IPF Team (2017) Tip failure and residual velocity in impact of hollow Al-6061 T6 projectiles on thin Al-6061 T6 plates. Procedia Eng 173:271–277
5. Iqbal MA, Gupta PK, Deore VS, Tak SK, Tiwari G, Gupta NK (2012) Effect of target span and configuration on ballistic limit. Int J Impact Eng 42:11–24 (2012).
6. Tanaka M, Moritaka Y, Akahoshi Y, Nakamura R, Yamori A, Sasaki S (2001) Development of a lightweight space debris shield using high strength fibers. Int J Impact Eng 26:761–772

9789811974731